Field Guide to the Rattan Palms of Africa

Terry C. H. Sunderland

Line drawings by
Lucy T. Smith

Kew Publishing
Royal Botanic Gardens, Kew

PLANTS PEOPLE
POSSIBILITIES

Dedicated to the memory of Paul Tuley (1927–2004) who did much to further our knowledge of African palms, and particularly rattans. He supported this work with immense enthusiasm and will be sorely missed.

All photography by the author except: *Eremosphatha cabrae* & *E. haullevilleana* by Paul Latham; *Laccosperma korupensis* by Joshua Linder; sheath of *Laccosperma acutiflorum*, fruits of *Eremospatha macrocarpa* by John Dransfield; leaf of *Eremospatha quinquecostulata* by Michael Balinga.

First published in 2007 by
Royal Botanic Gardens, Kew
Richmond, Surrey, TW9 3AB, UK
www.kew.org

ISBN 978-1-84246-180-8

British Library Cataloguing in Publication Data
A catalogue record for this book is available from the British Library

Production Editor: Michelle Payne
Typesetting and page layout: Margaret Newman
Design by Media Resources, Royal Botanic Gardens, Kew

Printed in the United Kingdom by CPI Antony Rowe, Eastbourne

For information or to purchase all Kew titles please visit
www.kewbooks.com or email publishing@kew.org

All proceeds go to support Kew's work in saving the world's plants for life

Contents

Acknowledgements

Much of the fieldwork for this book was funded by the Central African Regional Programme for the Environment (CARPE) and the United State Forest Service (USFS) Office for International Programmes. The International Network for Bamboo and Rattan (INBAR) funded the preparation of the taxonomic revision as part of the author's PhD study. Additional fieldwork and the preparation of extension manuals on rattan harvesting and cultivation were undertaken as part of a long term project by the African Rattan Research Programme and funded by the UK's Department for International Development (DFID) (Projects R7636 and R7636E).

Particular thanks are extended to Dr John Dransfield of the Royal Botanic Gardens, Kew, for initially identifying the need for basic work on African rattans and for providing considerable advice and encouragement during the preparation of this field guide. Thanks also to Lucy Smith for producing the excellent botanical illustrations.

A great many collaborators provided a huge amount of support and encouragement for this study and wider work on the promotion and development of African rattans. I would particularly like to thank my long term field assistant Dinga Njingum Franklin for his diligence in the field. The following people also contributed in some way to the preparation of this field guide: Gaston Achoundong, Charles Adu-Anning, Guillaume Akogo, Leeanne Alonso, Zachary Akum, William Baker, Vincent Beligne, Paul Blackmore, Dominic Blay, Henry Borobou Borobou, Mark Buccowich, Phil Burnham, Martin Cheek, Jim Comiskey, Stefan Cover, James Culverwell, Louis Defo, Emmanuel Ebanyele, Raphael Ebot, Martins Egot, Tom Evans, Barend van Gemerden, David Harris, William Hawthorn, Andrew Henderson, Ian Hunter, Daiou Joiris, David Kenfack, Paul Latham, Joshua Linder, Ruth Malleson, David Mbah, Tunde Morakinyo, Sainge Moses, Joseph Nkefor, Nouhou Ndam and the staff of the Limbe Botanic Garden, Ousseynou Ndoye, Crisantos Obama, Andrew Oteng-Amoako, John Palmer, Marc Parren, Jean-Pierre Profizi, Cherla Sastry, Tony Simons, Laurent Somé, Frank Stenmanns, Nicodeme Tchamou, Zach Tchoundjeu, Duncan Thomas, Theodore Trefon, Johan van Valkenberg, Paul Vantomme, Lee White and Zachary Nzooh.

Much gratitude is extended to the curators and directors of all of the herbaria who allowed me to visit their collections or who provided loan specimens. There are too many to list here but I am grateful to them all.

A special thank you also to all the chiefs, council members, harvesters, artisans, guides and other members of the communities I have worked with in the past few years (as part of the African Rattan Research Programme), who have provided hospitality, assistance, support, good humour and palm wine, often in copious quantities.

Heartfelt thanks are extended to my colleagues of the African Rattan Research Programme, notably Michael Balinga and Stella Asaha for all their work on the rattans of Africa over the past few years. We have been ably assisted by Mercy Abwe Dione, Mokabe Nanje Barnabas, Mboh Hyacinth and Eunice Njoh Fombod; many thanks to all of you and good luck with the launch of your new NGO, Forests Resources and People (FOREP).

Finally, a further dedication to Steve Gartlan, Anacletus Koufani, George Tenati and Mukete Wilfred: dear friends and colleagues who are no longer with us and whose wisdom and dedication are sorely missed.

Introduction

This field guide is intended to provide the necessary baseline information on taxonomy and utilisation of African rattans necessary for their improved management. Users of this book will be able to identify the rattans they encounter with relative ease through the use of descriptive keys and line drawings, together with photographs of the key features of each species. The fully illustrated species descriptions are primarily based on vegetative features which facilitate the identification of species in the field at any time of the year, but particularly when flowers and fruits are not being produced.

Field guides are based on understanding the way species relate to each other and how they are different. Taxonomic work of this sort is not purely an academic exercise; in this case it is an essential basis for the development and management of the rattan resource. It is important that the differences between species are clearly understood so that we know which species are of commercial importance and how they may be distinguished from other species which may not be utilised. This knowledge is essential in order to undertake meaningful inventories of commercially important species and to be able to assess the potential of each species for cultivation and sustainable management. A structured taxonomic framework also ensures that any experimental or development work undertaken is replicable.

What are rattans?

The word rattan is derived from the Malay 'rotan', the generic name for climbing palms. Rattans are climbing palms belonging to the Palm family (Palmae or Arecaceae). There are around 650 different species of rattan in 13 genera, concentrated solely in the Old World tropics of Africa and Asia. Rattans belong to the Calamoideae, a large subfamily of Palmae. All species within the Calamoideae are characterised by overlapping reflexed scales on the fruit and many of these climbing palms are very spiny.

Four genera of rattans, represented by 22 species, occur in West and Central Africa. In common with their Asian relatives, the rattans of Africa form an integral part of subsistence strategies for many rural populations as well as providing the basis of a thriving urban-based industry. Although many of the African rattan species are used locally for a multiplicity of purposes, the commercial trade concentrates on the bulk harvest of only a few widespread and relatively common species.

African rattans have long been recognised by donor agencies and national governments as having a potential role to play on the world market as well as a great role within the regional Non-Timber Forest Products (NTFP) sector of Africa. As increased interest is being shown in the potential role of high value NTFPs to contribute to conservation and development, rattan is frequently mentioned as a product that could be developed and promoted in a useful way. However, the development of the rattan resource in Africa has, until recently, been hindered by a lack of basic knowledge about the exact species used, their ecological requirements and the socio-economic context of their utilisation. To date it has not been possible to design appropriate management strategies that might be implemented to ensure their sustainable, and equitable, exploitation. Recent research has concentrated on the taxonomy, ecology and utilisation of these important species. This field guide is a direct output of that research and will provide a useful tool for scientists, foresters, students and extension workers to identify and sustainably manage the rattan species in their region of Africa.

Ecology and distribution

Rattans are widespread throughout West and Central Africa and are a common component of the forest flora. Some species, such as *Laccosperma secundiflorum* and *Eremospatha macrocarpa*, have large ranges and occur from Liberia to Angola, whilst *Calamus deërratus* is widely distributed and occurs from the Gambia, across to Kenya and southwards to Zambia. The greatest diversity of rattan species is found in the forests of Central Africa. Eighteen of the 22 known African rattan species occur in Cameroon alone.

Within this forest zone, rattans occur in a wide range of ecological conditions. The majority of species occur naturally in closed tropical forests and are early gap colonisers. Because of this, many taxa are extremely light demanding and respond well to a limited reduction in the forest canopy. Increases in forest disturbance, such as selective logging activity, encourage the regeneration of rattans and these palms are often a common feature along logging roads and skid trails. For some species, their light-demanding nature is such that they are often the earliest colonisers of disturbed areas. Other species of rattan, notably *Calamus deërratus*, grow in permanently or seasonally inundated forest or swamps, whilst other species, such as *Laccosperma opacum* and *L. laeve*, are highly shade-tolerant and prefer to grow under the forest canopy.

The seeds of most African rattans are dispersed predominantly by hornbills. However, a number of mammals, in particular the drill, mandrill, chimpanzees, gorillas and elephants are also key dispersal agents. The seeds are often scattered far from the mother plant. Interestingly, high germination rates also occur near to the parent plant through fruit fall, particularly in areas where over-hunting has led to a significant decline in animal dispersal agents and the fruits are not eaten.

Despite intensive field work and herbarium collection, especially in the past several years, no obvious pattern has been identified for flower development and seed production for most African rattan species.

The uses of rattan

The most important rattan product is cane; this is the rattan stem stripped of its leaf sheaths and outer skin, although this outer layer is also often used for weaving. The inner stem is solid, strong and uniform, yet highly flexible. The canes are either used whole, especially for furniture frames, or are split, peeled or cored for matting and basketry.

Local names

The extensive local naming systems for rattans often encountered in the field reflect the social significance of rattan. These local taxonomies have developed to reflect rattan as it grows wild in the forest, as well as how it is used. For example, a widespread species may be referred to by many names, as its range encompasses a number of dialectic and ethnic groups. Similarly, a single species can be given many names reflecting the different uses of the plant or the various stages of development (from juvenile to adult with very distinct morphological differences between the two). Commonly, local names for 'cane' are given to a wide range of species, even though they may be morphologically different and they may be used in different ways.

In some instances, species that have no use are often classified according to their 'relationship' to those that are used. These are often along kinship lines and species may be referred to as '*uncle-of...*' or '*small brother of...*' reflecting their perceived relationship and similarity to species that are widely utilised. In the past, serious confusion has arisen from the uncritical use of vernacular names and has contributed to the misconception that all species have commercial potential. Local names should be used in conjunction with classical taxonomic methods. However, local classifications often reflect subtle differences in morphology and utilisation that are not often noticed by the scientist. When a resource is as widely used and important to local people as rattan, vernacular and scientific names should be used with close reference to each other and not on a mutually exclusive basis.

Descriptive notes

The following defined terms are used in the keys and species descriptions to help the field worker trying to identify a rattan to species level.

The rattans of West and Central Africa are **clustered** – meaning each individual rattan has many stems. The stems are covered in **sheaths**; a key diagnostic feature. Mature sheaths in some taxa (*Eremospatha* and *Calamus*) bear a distinct **knee** directly below the leaf junction. In *Laccosperma* and *Oncocalamus* especially, there is an extension of the sheath beyond the insertion of the petiole – this is referred to as an **ocrea**. In *Laccosperma*, the ocrea is dry and often pointed. In *Oncocalamus* and *Eremospatha*, the ocrea is horizontal and green.

There are two different types of climbing whip: the **cirrus** is an extension of the leaf-tip armed with grapnel hooks called **acanthophylls**. *Oncocalamus*, *Laccosperma* and *Eremospatha* have this type of climbing mechanism. A long whip-like organ, the **flagellum** arises from the leaf sheath, but only occurs in *Calamus deëratus*. It is in fact a modified inflorescence and the flowers are produced along the flagellum.

Some species of *Oncocalamus* sometimes possess an undifferentiated petiole, in both mature and juvenile stems, that bears no leaflets or other leaf material only acanthophylls; this **elaminate rachis** can sometimes be as long as three metres.

Leaflets are arranged most commonly in a regular pattern although some, especially in the juvenile state, may be grouped and fanned, giving a **plumose** appearance (e.g. *Eremospatha quinquecostulata*), or completely **bifid** (arrow-head shaped) in the juveniles of some species (*Laccosperma*, *Eremospatha*, *Oncocalamus*). The leaflet shape is very variable, from linear to strongly diamond shaped. The apex of the leaflets may be somewhat raggedy, or **praemorse**, in appearance e.g. *Eremospatha wendlandiana*.

Inflorescences are produced in two main ways: in the first, several inflorescences are produced simultaneously in the axils of often reduced leaves at the stem tips, and after flowering and fruiting the stems die, usually replaced by sucker shoots at the base. This is known as hapaxanthy. *Laccosperma* is **hapaxanthic**. The second way is when stems reaching maturity go on producing one or more inflorescences often every year and the individual stem has the capability of unlimited growth. This is known as pleonanthy. *Calamus, Oncocalamus* and *Eremospatha* are all **pleonanthic**. *Eremospatha* and *Laccosperma* have male and female parts on the same flower (**hermaphroditic**). *Oncocalamus* has separate male and female flowers on the same inflorescence (**dioecious**) and *Calamus* has separate male and female flowers on different individuals or stems (**monoecious**).

Rattan fruits are covered in vertical rows of reflexed scales. A beak is often present at the tip of the fruit, tipped by the remains of the stigmas (most notable in *Laccosperma*); the remains of the petals and sepals are often found at the base of the fruit (*Laccosperma, Calamus*). Inside the **fruit**, there is usually a single seed, but some *Laccosperma* may have up to three seeds per fruit. The **seed** sometimes has a fleshy outer covering called the **sarcotesta** that may be sweet (*Laccosperma*) or bitter tasting (*Eremospatha*, *Calamus*).

Guide to collecting herbarium specimens

Rattans are notoriously difficult to collect and consequently are much under-represented in Herbaria. To collect rattans, it is best to work initially with known harvesters who are adept at pulling the long canes down from the canopy until you become familiar with the method used. Often it is far easier to collect specimens from gaps or roadsides where access is better. Because of the spines, you will need gloves!

One collection can be made up of five or six separate sheets, plus any associated collections such as DNA leaf material in silica gel and flower/fruiting material in spirit. Because of the large number of single components making up one rattan voucher collection, *it is important that each individual item is tagged and numbered, preferably using jewellers tags.*

In general, because the leaves are often very large, and because of the diagnostic importance of the leaf/stem junction and the cirrus, up to three representative portions should be collected and pressed in a separate sheet of newspaper.

The following plant parts are needed to make a voucher specimen:

1. The apex or the upper portion of the leaf, along with the cirrus (in full) which can be folded into a concertina shape or a circle prior to pressing.
2. A middle section of the leaf, revealing the size, shape, arrangement and distribution of the leaflets or subdivisions of the leaf.
3. The base of the leaf, showing the junction of the petiole and the stem. This is a very important diagnostic feature and is used to distinguish both genera and some species from one another. Often these sections are too large to be included in the press and can be dried separately and reconciled with the rest of the collection later on. For known commercial species, it is useful to also collect two additional stem sections: one with the outer covering and one that has been stripped to show the internal cane. For *Calamus*, the flagellum arising from the stem should be collected and can be folded much the same as the cirrus at the leaf apex (discussed above).
4. Representative portions of other important plant organs such as flowers and fruits. These can be placed either in the press and/or in a container with spirit (a combination of the two is preferred).

Making notes

Because palms are too big to be represented fully in a herbarium collection, a lot of information will still be missing after a visual inspection of a rattan voucher. Therefore extensive notes need to be taken on any distinctive aspects of the habit, stem, sheaths, leaves, inflorescences or fruits that are not evident in the collected specimen. Consider the following points when making notes:

1. A very important distinguishing feature of palms is whether they are solitary (single-stemmed) or clustered (multiple-stemmed), high climbing, erect, or acaulescent. Record the number of individual stems along with the maximum length of the stems. Measure and record stem diameter both with and without sheaths, internode length and number of branches (if any).
2. The overall length of the leaf (noting the length of the petiole with leaflets and the cirrus – if present – separately) should be recorded.
3. It is also useful to record the total number of leaflets on a representative leaf. If it varies, give an accurate range.
4. The number and general arrangement of leaves may be useful information.
5. The length and position (i.e. axillary, terminal) of all flowering and fruiting branches should also be recorded along with the number of partial inflorescences, the sex and arrangement of the flowers and bracts. Record also colour and scent. With fruits, note the colour and ripeness.
6. Record the presence of ants and other insects (collect in spirit if possible) and any other potentially useful ecological information.
7. It is also very important to record as accurate geographical and habitat information as possible; include political regions (Province, Division, etc.), and an accurate gazeteer (use a GPS if available) as well as noting the altitude, soil type, drainage, and surrounding vegetation (including whether the plant is found in a gap or below a closed canopy).

Key to the genera of African rattans

Rattans climbing by means of a cirrus, armed with acanthophylls or short, recurved thorns:

Leaf sheath without spines: stem sometimes with conspicuous knee below leaf junction; lowermost leaflets often swept back across stem; leaflets variously shaped, often with **praemorse** apex, or leaflets entire: *Eremospatha* (page 10)

Leaf sheath armed with conspicuous spines:

Ocrea triangular, not tubular, drying; spines on leaf sheath long, slender and sparsely to densely arranged, not easily detached; inflorescence produced simultaneously in the axils of the distal leaves; **hermaphroditic** flowers in dyads, rarely triads: *Laccosperma* (page 34)

Ocrea well-defined, tubular, not dry; spines on leaf sheath short, irregularly spaced, black or brown, triangular, easily sloughing off; inflorescence pendulous, produced in axils (**pleonanthic**); **monoecious** flowers in clusters of 7–11: *Oncocalamus* (page 50)

Rattans climbing by means of a flagellum emerging from the leaf sheath: *Calamus* (page 62)

Key to the species of *Eremospatha*

Stem with sheaths ± 1 cm in diameter; knee highly conspicuous or absent; leaflets ± regularly arranged, or inequidistant;

Knee present, conspicuous beneath leaf:

Ocrea entire, horizontally or obliquely truncate, or somewhat saddle-shaped:

Mature leaflets few in number (<20 on each side of the rachis) lowermost leaflets reduced, ± shaped as the mature leaflets; cirrus armed with reflexed spines:

Mature leaflets obovate-elliptic, inflorescence glabrous: *E. hookeri* (page 11)

Mature leaflets obovate to suborbicular; inflorescence profusely papillose to give brown velvety appearance: *E. cabrae* (page 14)

Mature leaflets many in number (>20 on each side of the rachis), linear-lanceolate to ovate to rhomboid, lowermost leaflets reduced, linear, strap-like; cirrus unarmed:

Sheath ± triangular in cross section, lowermost leaflets linear, strap-like; mature leaflets linear-lanceolate with finely acuminate apex: *E. laurentii* (page 16)

Sheath ± circular in cross section, lowermost leaflets linear to ovate; mature leaflets obovate-elliptic to oblanceolate to rhomboid, with distinctly **praemorse** apex: *E. dransfieldii* (page 18)

Ocrea longitudinally splitting into v-shape, or sometimes tattering;

Leaflets rhomboid or trapezoid with straight margins; cirrus armed with reflexed spines; inflorescence bracts inconspicuous, minute, <2 mm long: *E. wendlandiana* (page 20)

Leaflets linear-lanceolate; cirrus unarmed; inflorescence bracts conspicuous, up to 5 mm long: *E. barendii* (page 22)

Knee absent:

Leaflet apex narrowly to broadly **praemorse**:

Leaflets somewhat papery, mid-green; spines on leaflet margin always forward-facing:

Leaflets opposite to subopposite, linear-lanceolate, apex narrowly **praemorse**; ocrea with rounded lobe adaxial to the leaf; cirrus unarmed: *E. macrocarpa* (page 24)

Leaflets subopposite to alternate, cuneate, spathulate or ovate, apex moderately to strongly **praemorse**; ocrea obliquely truncate; cirrus armed: *E. haullevilleana* (page 26)

Leaflets somewhat leathery, grey-green; spines on leaflet margin reflexed and both forward and rear-facing: *E. tessmanniana* (page 28)

Leaflet apex entire, terminating in a conspicuous apiculum: *E. cuspidata* (page 30)

Stem with sheaths <1 cm in diameter; knee inconspicuous, linear, ridge-like; leaflets conspicuously inequidistant, in groups, clustered or somewhat **plumose**: *E. quinquecostulata* (page 32)

Eremospatha hookeri

(G. Mann & H. Wendl.) H. Wendl.

Les Palmiers 244 (1878)

Joseph D. Hooker (1817–1911), botanist and former Director of the Royal Botanic Gardens, Kew

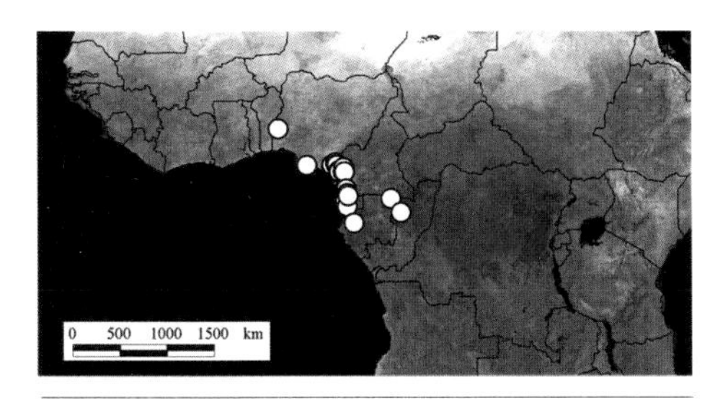

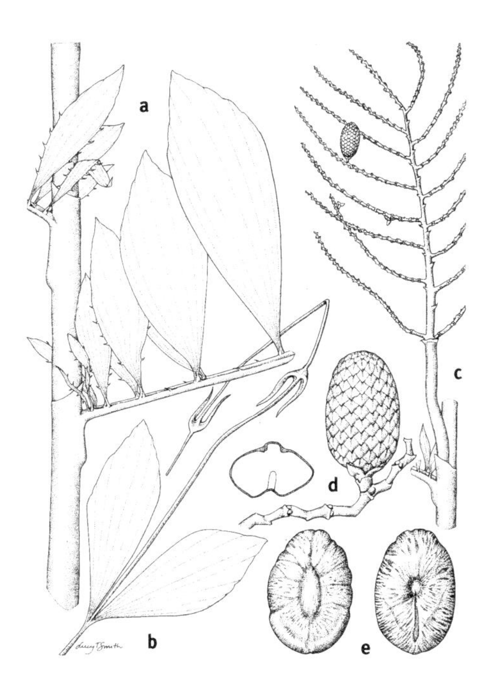

a. Sheath and leaflets; **b**. Leaflets and cirrus; **c**. Infructescence; **d**. Fruit; **e**. Seed (2 views).

Description

A relatively slender rattan with a conspicuous knee and ovate-oblanceolate leaflets. **Clustering** rattan palm. **Stems** up to 30 m long, 15–20 mm in diameter. **Sheath** 20–30 mm in diameter, unarmed, striate; **ocrea** entire, or obliquely truncate, often drying grey-brown, with distinct linear ridge on the adaxial side; **knee** linear, 1.5–3 cm long, somewhat abrupt at base. **Juvenile leaves** shortly petiolate, roundly **bifid**, soon becoming pinnate. **Adult leaves** sessile, up to 1.5 m long; **leaflets** oblanceolate to almost rhomboid, broadly **praemorse** at apex, lowermost leaflets smaller than the rest, linear to ovoid reflexed and laxly clasping the stem; **cirrus** 50–70 cm long; **acanthophylls** 2–2.5 cm long. **Inflorescence** produced in axils, glabrous, up to 40 cm long. **Flowers** not known. **Fruit** 1–2 seeded, ovoid to cylindrical; **seed** compressed, flattened on one side with somewhat wavy margins.

Uses

The split cane is used as a strong binding material. In Nigeria, the base of the leaf sheath is used as a chewstick.

Conservation status

Not threatened.

Habitat Particularly shade tolerant and more commonly found under a forest canopy.

Distribution Currently this species has a distinct Guineo-Congolian distribution.

Eremospatha hookeri

Juvenile in cultivation

Aerial branching

Mature stem

Leaf sheath showing knee

Leaf sheaths

Vernacular names

Nigeria: *epa-emele* (Yoruba); *inima ború* (Ijo-Izon); *itomi* (Ekit). **Cameroon:** *ki-yince* (Balundu-Bima); *mbunden* (Bakundu-Balue). **Equatorial Guinea:** *alua-nlong* (Fang). **Gabon:** *gigorula* (Sira).

Eremospatha cabrae

(De. Wild. & Th. Dur.) De Wild.

Annals du Musée Congo Belge 5(1): 95 (1904)

Captain E. Cabra, Belgian administrator and explorer

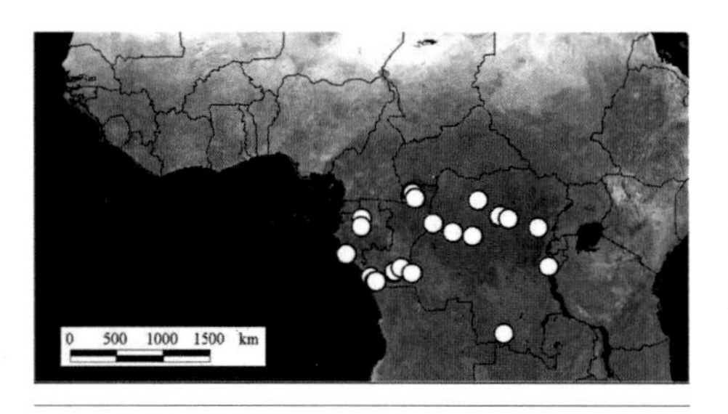

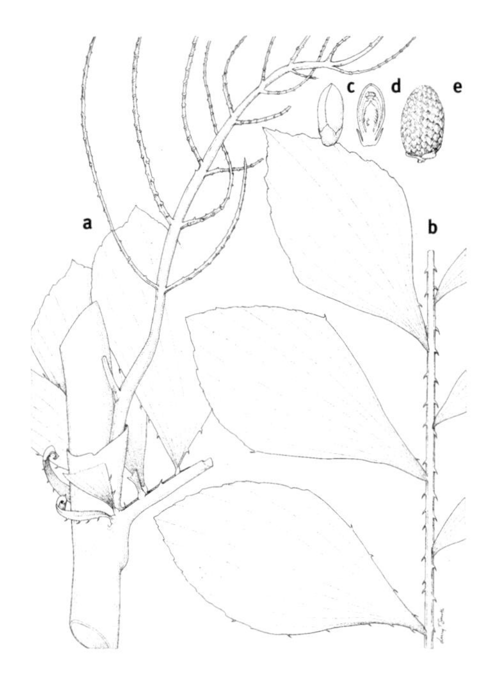

a. Sheath & inflorescence; **b.** Leaflets; **c.** Flower; **d.** Flower section; **e.** Fruit.

Description

A slender to moderate rattan distinguished by the possession of conspicuously rounded leaflets and a velvet-like covering on the inflorescence. **Clustering** rattan palm. **Stems** up to 50 m long, more commonly 20–30 m, 10–15 mm in diameter. **Sheath** up to 25 mm in diameter, unarmed, striate; **ocrea** entire, saddle-shaped with 1–1.5 cm lobe extending beyond leaf junction; **knee** narrow, linear, 2.5 cm long, abrupt at the base. **Leaves** sessile, up to 1 m long; **leaflets** obovate to trapeziform, broadly **praemorse** at apex, ciliate spiny, lowermost leaflets smaller than the rest, erect or reflexed and laxly clasping the stem; **cirrus** up to 1 m long; **acanthophylls**, slender, up to 3.5 cm long. **Inflorescence** produced in axils, covered with velvety covering, up to 40 cm long. **Flowers** borne in close pairs. **Fruit** 1-seeded, cylindrical to rhomboid; **seed** flattened on one side with a shallow depression.

Uses

Split stems are used for the fabrication of temporary market baskets and other woven products or are used complete for the manufacture of furniture frames, particularly in the absence of better quality cane species. Whole stems are also employed for the construction of cane bridges. The base of the leaf sheath is also used as a toothbrush.

Conservation status

Not threatened.

Habitat More commonly encountered in swampy areas than terra firma forest.

Distribution Restricted to Gabon, southwards to northern Angola and across the lowland forests of the Congo Basin.

Eremospatha cabrae

Eremospatha cabrae

Type specimen of *E. cabrae*

Vernacular names

Gabon: *osono* (Tsogo); *osono* (Pinji); *ozono* (Myene); *li-bamba* (Vili); *nkolé* (Kélé); *nkolu* (Seki); *du-bamba* (Barama); *du-bamba* (Lumbu); *ivéta* (Duma); *iló-lóngo* (Kota); *u-lóngo* (Benga); *lé-mbumu* (Ndumu); *nlong* (Fang). **DR Congo:** *li-findo* (Lombo); *lu-bambi* (Kituba); *e-safa* (Mongo-Nkundu); *ki-sakata* (Kete). **Angola:** *m'bamba* (Mbundu-Luanda).

Eremospatha laurentii

De Wild.

Bulletin du Jardin Botanique de l'Etat à Bruxelles 5: 147 (1916)
Marcel Laurent (1879–1924), Belgian botanist

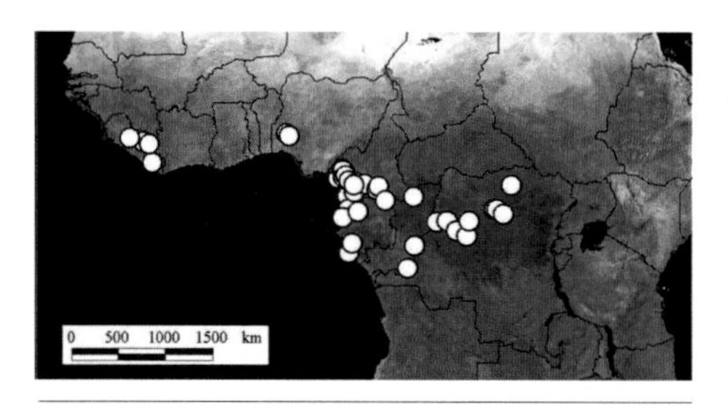

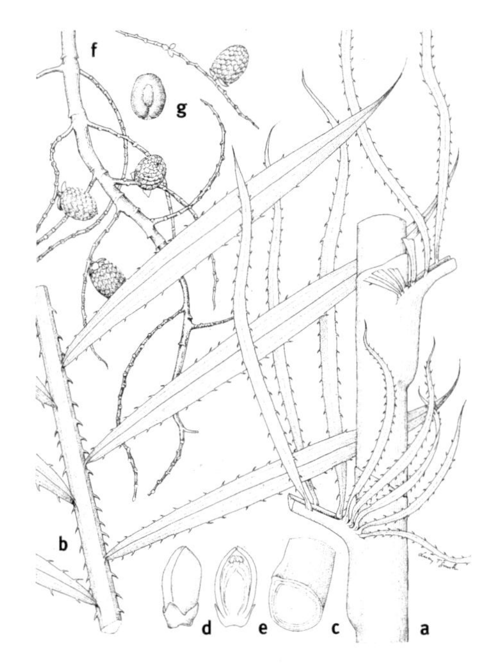

a. Sheath; **b.** Stem section; **c.** Leaf; **d.** Flower; **e.** Flower section; **f.** Infructescence section; **g.** Seed.

Description

A robust rattan easily recognisable by the conspicuous knee, the strap-like linear lowermost leaflets and the entire leaflet apex. **Clustering** robust rattan palm. **Stems** somewhat triangular in cross-section, up to 30 m long, 18–25 mm in diameter. **Sheath** 25–30 mm in diameter, unarmed, lightly striate; **ocrea** entire, obliquely truncate, extending to 2 cm; **knee** highly conspicuous, narrow, linear, 50–80 mm long, abrupt at the base. **Leaves** sessile, up to 1.2–1.5 m long; **leaflets** linear-lanceolate to ovate, finely acuminate at apex which often breaks off to give blunted appearance, lowermost leaflets smaller than the rest, linear, strap-like, armed along margins with bulbous-based spines, laxly swept back across stem or tightly clasping; **cirrus** 1.2–1.5 m long; **acanthophylls** robust, 3–4 cm long. **Inflorescence** produced in axils, glabrous, up to 35 cm long, flattened in cross section, erect, arching. **Flowers** borne in pairs. **Fruit** 1-seeded, globose or cylindrical; **seed** compressed, rounded on one side.

Uses

This species is rarely used for either furniture manufacture or basketry as the cane is of poor quality.

Conservation status

Not threatened.

Habitat Found in both open areas as well as in closed-canopy forest. Responds particularly well to selective logging and is a common component of regrowth vegetation where it occurs.

Distribution Occurs predominantly in the lowland forests of the northern Congo Basin. However, there are outliers of this species found in the forests of Upper Guinea, including Nigeria and Sierra Leone, with a pronounced disjunction from Côte d'Ivoire to Benin.

Eremospatha laurentii

Habit

Leaflets

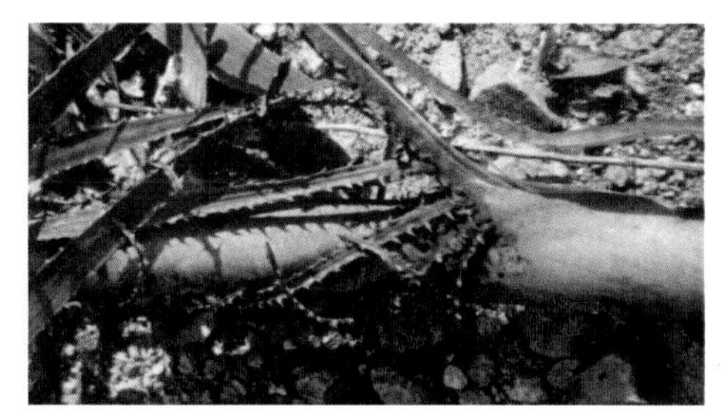

Leaf sheath showing knee

Adult stem

Vernacular names

Sierra Leone: *bongei* (Mende). **Cameroon:** *kpakpa* (Ewondo). **Central African Republic:** *bo-kondi* (Banda-Yangere). **Equatorial Guinea:** *ebuat* (Fang). **DR Congo:** *bo-ngale* (Mongo-Nkundu); *ikonga* (Lombo); *nkelele mo-none* (Lingala); *nkoli* (Bali).

Eremospatha dransfieldii

Sunderl.

Kew Bulletin 58: 987–990 (2003)

John Dransfield, Palm Specialist, Royal Botanic Gardens, Kew

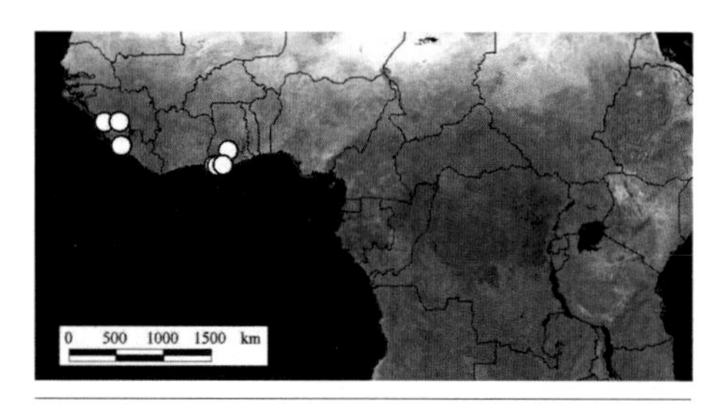

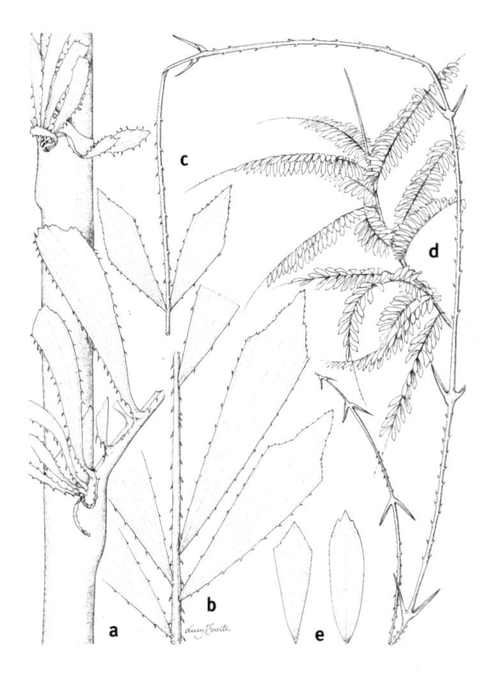

a. Sheath; **b.** Adult leaflets; **c.** Cirrus; **d.** Habit; **e.** Leaflets.

Description

A moderate to robust rattan recognisable by the presence of a conspicuous knee and leaflets that are highly variable in shape, from linear to somewhat rhomboid. The sheath is often profusely covered in colonies of scale insects. **Clustering** moderate to robust rattan palm. **Stems** up to 40–50 m long, 18–24 mm in diameter. **Sheath** 25–30 mm in diameter, unarmed, lightly striate, often covered in scale insects; **ocrea** entire, obliquely truncate, extending 1–2 cm above the leaf junction; **knee** conspicuous, linear, up to 2–4 cm long. **Leaves** sessile, up to 3.5 m long; **leaflets** highly variable in shape, obovate-elliptic to oblanceolate to rhomboid, with distinctly **praemorse** apex; lowermost leaflets smaller than the rest, linear (strap-like) or broadly lanceolate, reflexed and laxly or tightly clasping the stem; **cirrus** 1.5–2 m long; **acanthophylls** 3–4 cm long. **Inflorescence** unknown. **Flowers** unknown. **Fruit** unknown.

Uses

Used mainly for furniture frames and coarse basketry and traded in Ghana and Côte d'Ivoire.

Conservation status

Vulnerable due to limited distribution and over-harvesting to supply the growing cane trade, particularly in Ghana.

Habitat This species is extremely light demanding, occurring naturally in gap vegetation and forest margins.

Distribution Very localised throughout its range occurring only in moist evergreen forest with rainfall >2000 mm a year. The main distribution is centred in the Western region of Ghana and eastern Côte d'Ivoire with additional populations in Sierra Leone.

Eremospatha dransfieldii

Juvenile in forest

Habit

Leaflets showing variation

Mature sheath

Leaflets showing variation

Vernacular names

Sierra Leone: *balu* (Kono); *mbalu* (def. *-uí*) (Mende); *ra-thamp* (Themne). **Côte d'Ivoire:** *kpè-pun* (Attié); *tami* (Dioula); *kou gnain* (Gouro); *niböi-gain* (Wè). **Ghana:** *mfia* (Twi).

Eremospatha wendlandiana

Dammer ex Becc.

Webbia 3: 290 (1910)

Hermann Wendland (1825–1903) German palm botanist and horticulturist

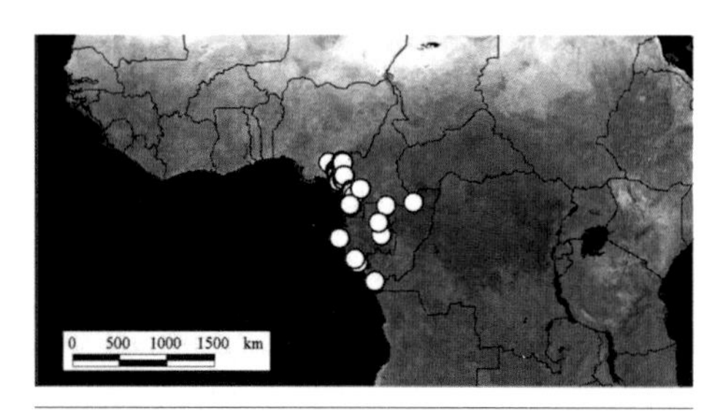

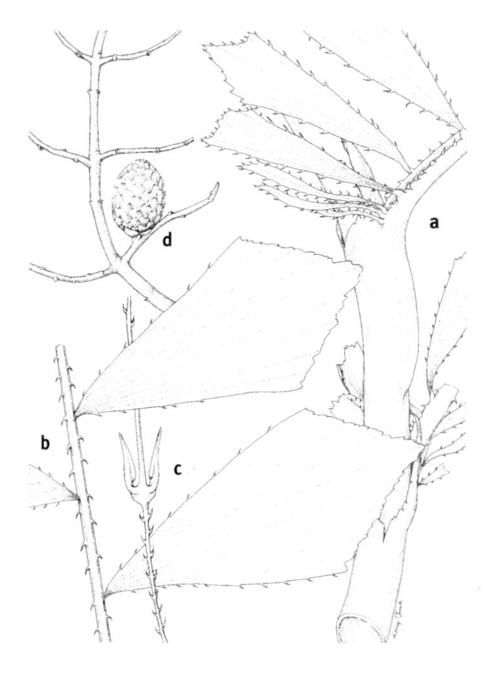

a. Sheath; **b.** Leaflets; **c.** Acanthophylls; **d.** Fruit on infructescence.

Description

A highly distinctive robust rattan with triangular rhomboid leaflets and a characteristically tattering ocrea. **Clustering** robust rattan palm. **Stems** up to 60 m long, 12–20 mm in diameter. **Sheath** 15–30 mm in diameter, unarmed, very lightly striate; **ocrea** drying brown and splitting longitudinally, sometimes with horizontal linear wrinkle opposite the leaf; **knee** highly conspicuous, narrowly linear, extending to 80 mm long, tapering at the base. **Leaves** sessile, up to 2 m long; **leaflets** strictly rhomboid or trapezoid with straight margins, broadly **praemorse** at apex, somewhat ciliate, spiny, very variable in size; **cirrus** up to 2 m long; **acanthophylls** slender 1.5–2 cm long. **Inflorescence** produced in axils, glabrous, up to 80 cm long. **Flowers** borne in pairs. **Fruit** 1-seeded, ovoid to cylindrical; **seed** compressed, flattened on one side with somewhat wavy margins.

Uses

Split stems are used to tie bamboo and stick-framed houses prior to plastering with clay, as well as for coarse basketry. The stem epidermis is also split and used for tying yams. The base of the leaf sheath is used as a chewing stick and the apex of the young emerging stems are often roasted and consumed by hunters. This species does not have much commercial value and is rarely traded.

Conservation status

Not threatened.

Habitat A common component of gap vegetation and forest margins although also present in closed-canopy forest.

Distribution From SE Nigeria to Gabon, commonly in coastal forest, although with outliers present in the swamp forests of the Central African Republic.

Eremospatha wendlandiana

Juvenile in cultivation

Habit

Leaflet

Leaf sheath with knee

Vernacular names

Nigeria: *eghounka* (Ekit). **Cameroon:** *cane basket* (Pidgin); *mua-echié* (Denya). **Equatorial Guinea:** *akot* (Fang). **Gabon:** *égoo* (Tsogo); *ngundju* (Punu); *ngundju* (Vumbu). **Congo:** *ma-bulu* (Téké).

Eremospatha barendii

Sunderl.

Journal of Bamboo and Rattan 1(4): 361–369 (2002)
Barend van Gemerden, Forestry Researcher (1967–)

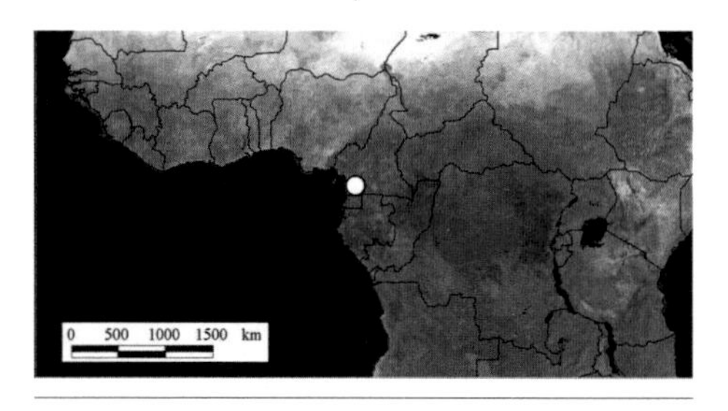

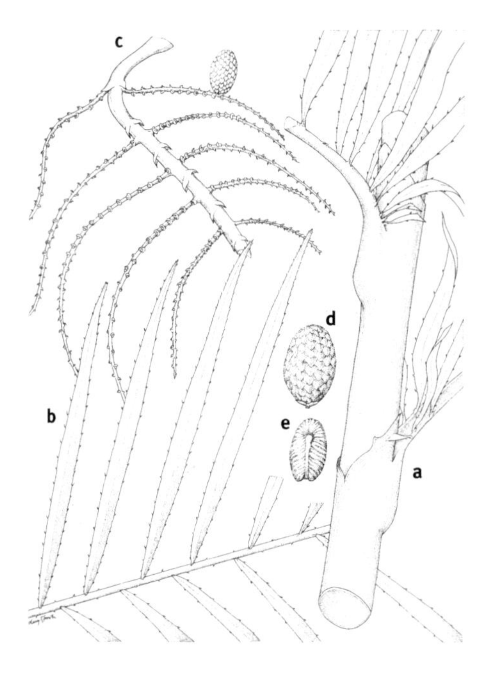

a. Sheath; **b.** Leaflets; **c.** Infructescence; **d.** Fruit; **e.** Seed.

Description

An uncommon species characterised by the presence of conspicuous bracts on the inflorescence. **Clustering** robust rattan palm. **Stems** up to 25–30 m long, *c.* 15 mm in diameter. **Sheath** *c.* 25 mm in diameter, unarmed, longitudinally striate; **ocrea** drying brown and splitting to form conspicuous v-shape on abaxial side; knee linear, 3–3.5 cm long, somewhat abrupt at base. **Leaves** sessile, up to 1.2 m long; **leaflets** linear-lanceolate, lowermost leaflets smaller than the rest linear-ovate, erect, or laxly reflexed across stem; **cirrus** 1.2 m long; **acanthophylls** 2.5 cm long. **Inflorescence** produced in axils, velvety, up to 30 cm long with conspicuous rachis bracts. **Flowers** not known. **Fruit** 1-seeded, broadly cylindrical; **seed** compressed, flattened on one side.

Uses

None recorded.

Conservation status

Endangered (known only from a single collection).

Habitat and **Distribution** A very poorly-known species, known from a single collection in a timber concession near Lolodorf, Cameroon that has recently been logged. The individual was found in a light gap in high forest.

Vernacular names

None recorded.

Rattan harvesting camp, Mokoko, Cameroon

Harvesting in the Southern Bakundu Forest Reserve, Cameroon

Cut canes of *L. secundiflorum* awaiting transportation, Mokoko, Cameroon

Canes of *L. robustum* drying, Bata, Equatorial Guinea

Rattan harvester, Rumpi Hills, Cameroon

Eremospatha macrocarpa

(G. Mann & H. Wendl.) H. Wendl.

Les Palmiers 244 (1878)
(Latin) "large-fruits"

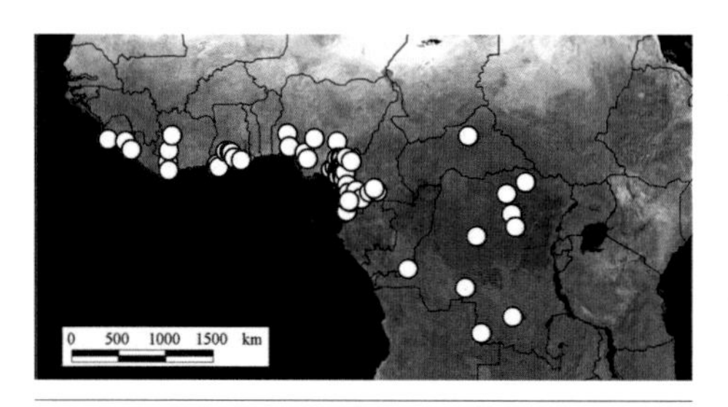

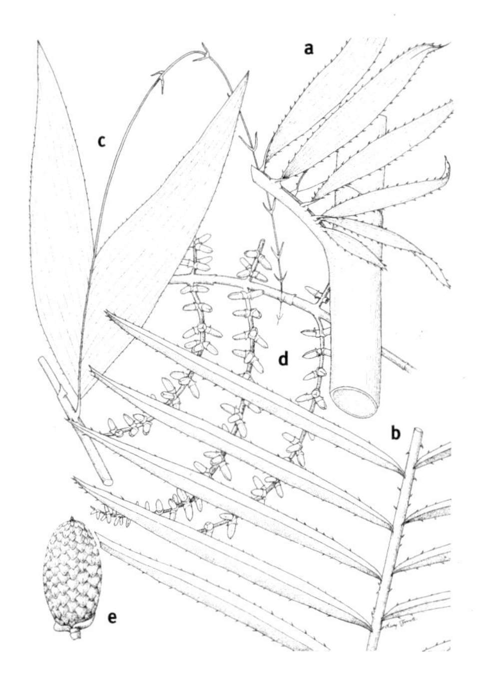

a. Sheath; **b.** Adult leaflets; **c.** Juvenile leaf; **d.** Infloresence; **e.** Fruit.

Description

A slender to moderate rattan easily recognisable by the absence of a knee, linear-lanceolate leaflets, with a narrowly **praemorse** apex. The leaves are strongly **bifid** when juvenile. **Clustering** slender to moderate rattan palm. **Stems** up to 50–75 m long, rarely to 150 m, 10–18 mm in diameter. **Sheath** 22–30 mm in diameter, unarmed, longitudinally striate; **ocrea** entire, saddle-shaped, with 2.5–4 cm long rounded lobe, often with linear ridge adaxial to the leaf, particularly on juvenile stems; **knee** absent. **Juvenile leaves** shortly petiolate, sharply **bifid. Adult leaves** sessile, 1–1.5 m long; **leaflets** linear-lanceolate, narrowly **praemorse** at apex, lowermost leaflets smaller than the rest, linear-ovate, reflexed and laxly clasping the stem; **cirrus** 1.2–2 m long; **acanthophylls** *c.* 3 cm long. **Inflorescence** produced in axils, glabrous, up to 55 cm long, arching outwards. **Flowers** in close pairs. **Fruit** 1- (rarely 2-) seeded, cylindrical; **seed** compressed, flattened on one side or with a shallow depression.

Uses

Reputed to be the best source of cane in Africa and of comparable quality to the small-diameter canes of SE Asia. Throughout its range it is widely traded and used for furniture construction, basketry, weaving and tying. Its long flexible stems also make it ideal for the construction of cane bridges. The powdered root is taken as a medicine for the treatment of syphilis in Ghana and Nigeria.

Conservation status

Not threatened, although becoming locally scarce due to over-harvesting.

Habitat This species is extremely light demanding, occurring naturally in gap vegetation and forest margins. As a result, it responds very well to selective logging and is a common component of regrowth vegetation.

Distribution Widespread and common, distributed from Sènègal in West Africa through to the lowland forests of the Congo Basin.

Habit

Juvenile in cultivation (note bifid leaves)

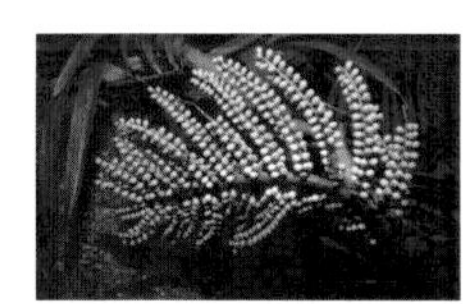

Inflorescence

Infructescence

Juvenile sheath

Mature sheath (note wrinkle)

Vernacular names

Sierra Leone: *penden* (Kissi); *balu* (Kono); *mbalu* (Loko); *mbalu, koto mbalu* = juvenile (Mende); *ra-thamp* (Themne). **Liberia:** *bìlì de bìlì* (Mano). **Côte d'Ivoire:** *ailè-mlé* (Anyin). **Ghana:** *mfia* (Akan-Asanti); *néné* (Nzima). **Benin:** *dekon* (Defi); *dekun vovo* (Gun-Gbe). **Nigeria:** *íkan* (Edo); *odu-aòa (Igbo); bórú* (Ijo-Izon); *ukan* (Yoruba); *ekakieri* = male (i.e. with no fruits), *irrumka* = female (with fruits) (Ekit); *iro* (Esan). **Cameroon:** *filet* (Trade); *cane rope* (Pidgin); *echié* (Denya); *nlong* (indef.) *melong* (def.) (Bulu); *bana ndongo* = young cane (*bana* = child) (Balundu-Bima); *nloun* (Baasa). **Equatorial Guinea:** *nlong* (indef.) *mi-long* (def.) = juvenile stems, *ongam* = adult (Fang). **Gabon:** *ke-gèma* (Lumbu); *nyèvila* (Sira); *ongam* (Fang); *ndètèse* (Kota); *iganga-tsungu* (Punu); *songu* (Vumbu); *tongo* (Tsogo); *mbubi* (Ndumu).

Eremospatha haullevilleana

De Wild.

Annals du Musée Congo Belge 5(1): 96 (1904)

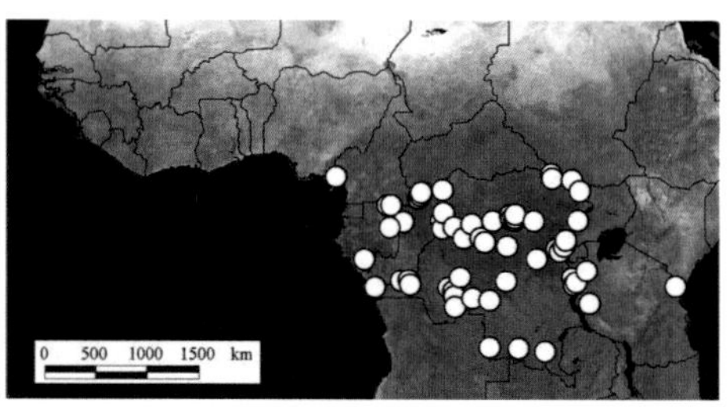

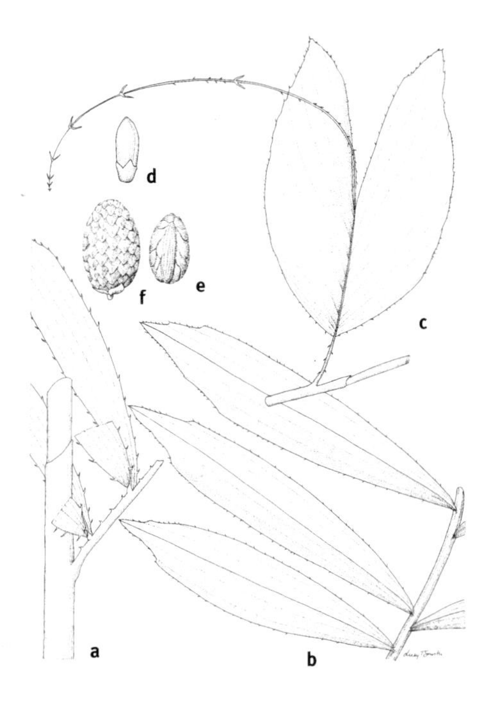

a. Stem; **b.** Leaflets; **c.** Juvenile leaf; **d.** Flower; **e.** Fruit; **f.** Seed.

Description

A slender to moderate rattan distinguished by a conspicuously striate leaf sheath and highly variable leaflets. **Clustering** rattan palm. **Stems** up to 25 m long, 6–15 mm in diameter. **Sheath** 10–25 mm in diameter, unarmed, conspicuously striate, often with sparse black indumentum; **ocrea** entire, obliquely truncate, extending 3–4 cm beyond leaf junction; **knee** absent. **Juvenile leaves** with petiole up to 15 cm long, leave **bifid**, with rather rounded lobes; **adult leaves** sessile, up to 1.2 m long; **leaflets** spathulate or ovate with uneven moderately to strongly **praemorse**; lowermost leaflets smaller than the rest, sometimes reflexed and laxly clasping the stem; **cirrus** up to 1 m long; **acanthophylls**, slender, up to 3.5 cm long. **Inflorescence** produced in axils, covered with velvety covering, 40–60 cm long. **Flowers** borne in close pairs. **Fruit** 1-seeded, ovoid to almost cylindrical; **seed** flattened on one side with slightly undulate margins.

Uses

This is a preferred species for basketry, weaving and furniture manufacture throughout its range. The stems are used whole for a wide range of products including for use as cables for cane bridges, furniture framework and building frames. The split stems are used for the fabrication of fish traps, noose-type snares to catch small terrestrial mammals and for the handrails of river bridges. The apical bud of this species is widely consumed. In DR Congo, the fruits are used for decoration, particularly in the manufacture of traditional collars, and the acanthophylls are used as fish hooks.

Conservation status

Not threatened.

Habit

Habitat Found in closed-canopy forest and in open areas. In common with *E. macrocarpa*, it is a species of terra firma forest and is not associated with swamp vegetation.

Distribution Restricted to the lowland forests of the Congo Basin. Unlike the majority of the rattan species, it is curiously absent from the coastal forest regions.

Habit

Vernacular names

Central African Republic: *pongbo* (Ngombe). **Congo:** *mbaama* (Téké). **DR Congo:** *li-findo* (Lombo); *mbowe* (Zande); *lu-popi* ((Nandi)); *n'kele* (Bangala); *m'bio* (Bangi); *lo-koli* (Kele); *ke-kele* (Lingala); *lu-kodi* (Luba-Shari); *lu-busi* (Tembo); *lu-bubi* (Lega-Mwenga); *yofoko* (Mungo-Nkundu); *lo-keko* (Lusengo); *kodi* (Luba-Kasai); *tukpuru* (Bhele). **Uganda:** *bibbobbi* (Amba); *enga* (Luganda). **Tanzania:** *urugage* (Ha).

Eremospatha tessmanniana

Becc.

Webbia 3: 278 (1910)

Günther Tessmann (born ? – 1926) German botanist and anthropologist

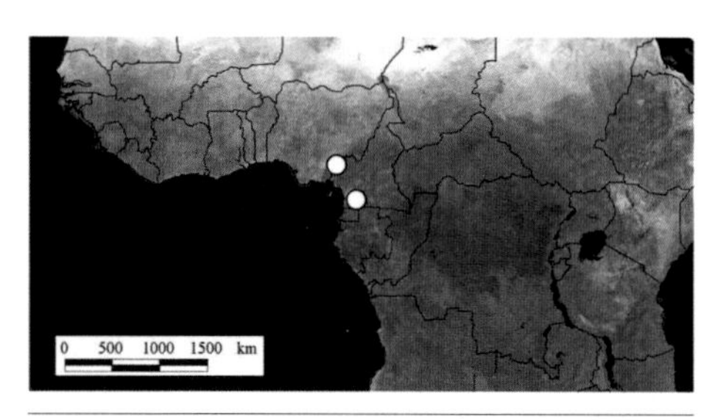

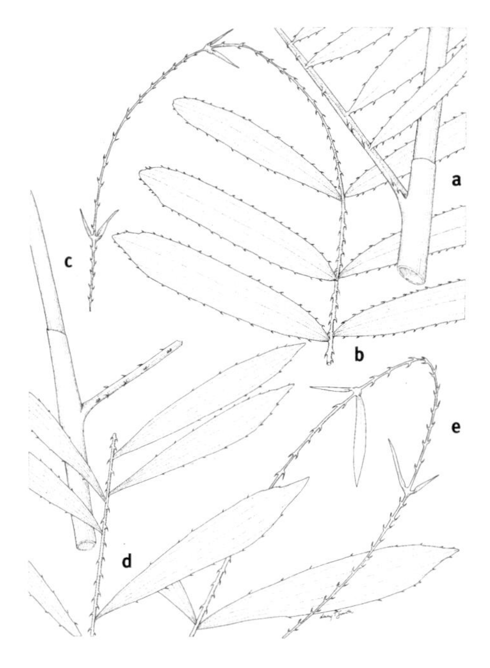

a. Sheath; **b.** Leaflets (**c. – e.** refer to *E. quinquecostulata*).

Description

A slender species of rattan characterised by rather glaucous grey-green leaflets and conspicuous forward and rear facing spines on the leaflet margins. **Clustering** slender rattan palm. **Stems** often branched, 60–80 m long, less commonly to 150 m, 10–12 mm in diameter. **Sheath** 12–15 mm in diameter, unarmed, longitudinally striate; **ocrea** entire, horizontally truncate, extending to 1.5 cm; **knee** absent. **Juvenile leaves** petiolate, **bifid**, with somewhat rounded lobes. **Adult leaves** sessile, or very shortly petiolate, up to 80 cm long; **leaflets** somewhat plumose, linear-elongate to lanceolate, attenuate at the base, rounded **praemorse** at apex, armed along margins with black-tipped spines, reflexed and reverse-facing at base and forward facing at apex; lowermost leaflets, smaller than the rest, lax, not reflexed; **cirrus** 40–60 cm long; **acanthophylls** slender, 2–2.5 cm long. **Inflorescence** unknown. **Flowers** unknown. **Fruit** unknown.

Uses

None recorded.

Conservation status

Vulnerable.

Habitat Found on well-drained soil in closed-canopy forest.

Distribution A relatively uncommon species of rattan known from only three localities: the Takamanda region of the Cameroon/Nigeria border, the cross border region of Cameroon, and the Rio Muni territory of Equatorial Guinea. Further collections might link this disjunction.

Mature stem

Seedlings on forest floor

Leaf

Leaf sheath

Reproduction through stolon development

Leaflet

Vernacular names

Cameroon: *calumé echié* (Denya). **Equatorial Guinea:** *ongam-akot* (Fang).

Eremospatha cuspidata

(G. Mann & H. Wendl.) H. Wendl.

Les Palmiers 244: (1878)

(Latin) refers to the finely pointed (apiculate) leaflet apex

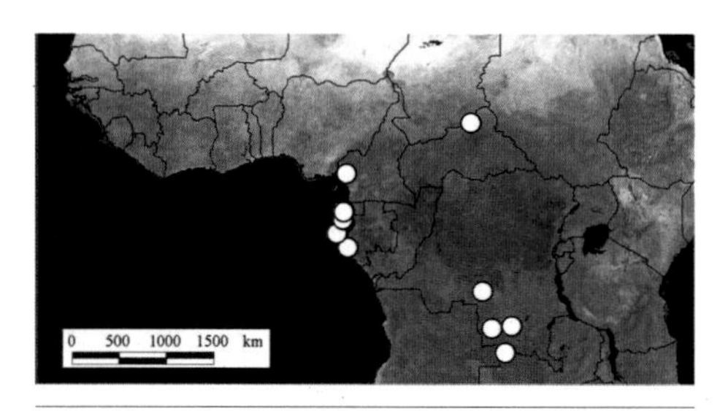

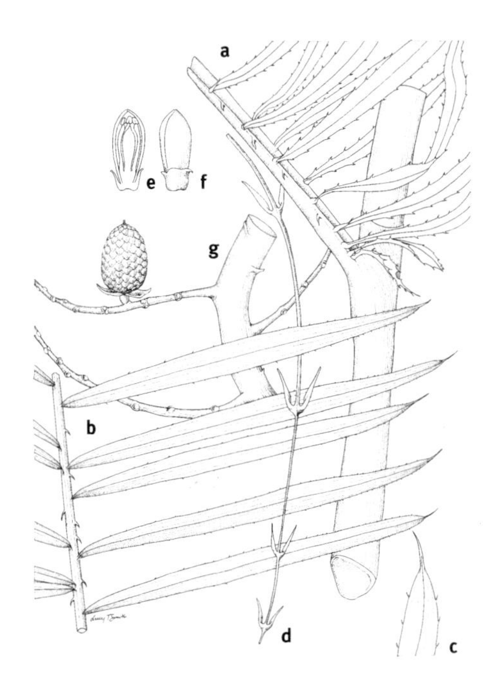

a. Sheath; **b.** Leaflets; **c.** Leaflet apex; **d.** Acanthophylls; **e.** Flower section; **f.** Flower; **g.** Fruit and infructescence.

Description

A relatively rare species of rattan characterised by the conspicuously apiculate leaflet apex. **Clustering** slender rattan palm. **Stems** 12–15 m long, 10–15 mm in diameter. **Sheath** 16–25 mm in diameter, unarmed, conspicuously longitudinally striate; **ocrea** entire, obliquely truncate, extending to *c.* 1 cm; **knee** absent. **Leaves** sessile, 1–1.3 m long; **leaflets** linear-lanceolate contracted at the base with a fine apiculum at the apex; lowermost leaflets smaller than the rest, linear-ovate, reflexed and laxly clasping the stem; **cirrus** 5–75 cm long; **acanthophylls**, in pairs, up to 3 cm long. **Inflorescence** produced in axils, glabrous, up to 30–40 cm long, erect or horizontal. **Flowers** in close pairs, very sweetly scented at anthesis. **Fruit** 1-seeded, cylindrical; **seed** flattened on one side with linear depression.

Uses

The split stems are used for light basketry and weaving, particularly in the absence of other species.

Conservation status

Not threatened.

Habitat *E. cuspidata* is highly unusual amongst the rattans of Africa in that it is most commonly found in the deep white sand savannah areas characteristic of the coastal forests of the Congo Basin, where it forms dense, scrambling thickets. However, *E. cuspidata* has also been encountered in gap vegetation in forest.

Distribution This species is relatively uncommon and is restricted to the forest areas of the Congo Basin.

Eremospatha cuspidata

Inflorescence

Leaflets (note pointed apex)

Habit

Mature stem

Infructescence

Vernacular names

Equatorial Guinea: *ndera* (Fang).

Eremospatha quinquecostulata

Becc.

Webbia 3: 279 (1910)
(Latin) "five main veins"

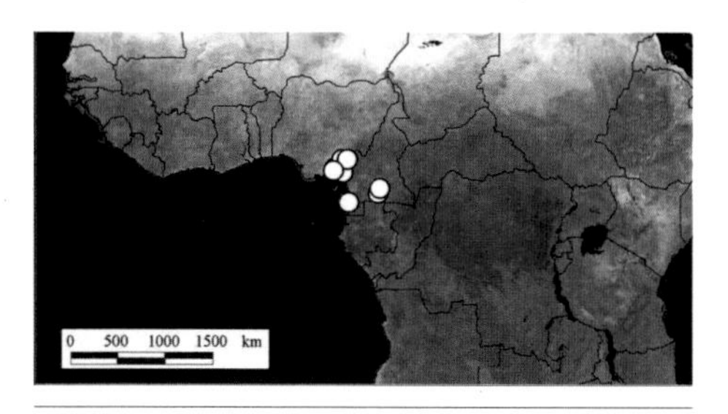

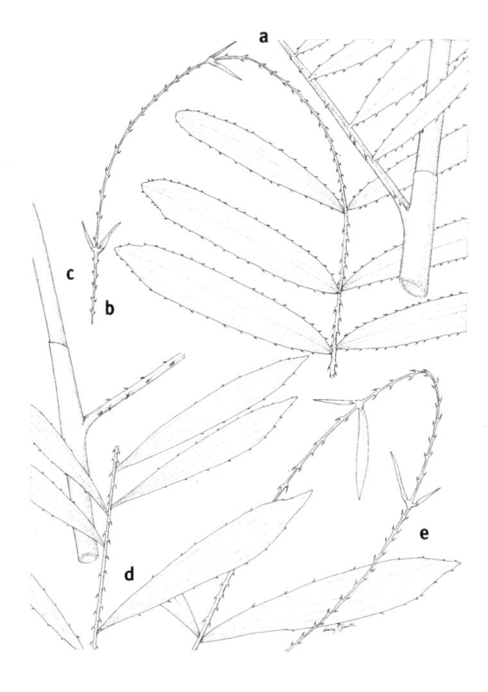

(**a.** and **b.** refer to *E. tessmanniana*); **c.** Sheath; **d.** Leaflets; **e.** Leaflets & cirrus.

Description

A slender species of rattan with leaflets clustered in groups of 4–6 and conspicuous leaflet veins. **Clustering** slender rattan palm. **Stems** 10–15 m long, 4–9 mm in diameter. **Sheath** 5–10 mm in diameter, unarmed, longitudinally striate; **ocrea** entire, obliquely truncate, extending to 1.2–1.7 cm; **knee** inconspicuous, ridge-like, vertically linear, tapering at base. **Leaves** sessile, or petiolate, 60–80 cm long; **leaflets** grouped in pairs or in 4's, irregularly clustered or somewhat plumose, lanceolate to loosely rhomboid, entire and acuminate to irregularly **praemorse** at apex; lowermost leaflets, slightly smaller than the rest; **cirrus** 40–50 cm long; **acanthophylls** slender, 2 cm long. **Flowers** unknown. **Fruit** unknown.

Uses

Not used in Cameroon due to the presence of other, more desirable, species of rattan. In Gabon the stem is split and employed for basic weaving, particularly in the absence of other species.

Conservation status

Vulnerable.

Habitat *E. quinquecostulata* is a slender rattan found only in high forest.

Distribution This species is known only from Cameroon and SE Nigeria.

Leaf

Leaflets

Leaf sheath

Habit

Vernacular names

Cameroon: *calumé-echié* (Denya). **Gabon:** *di-bula* (Sira).

Key to the species of *Laccosperma*

Canes slender, stems with sheaths 2–3.5 cm in diameter, with 10–12 leaflets on each side of the rachis; common in forest understorey:

Cirrus armed with conspicuous acanthophylls, leaflets ovate:

Leaflet margin armed with forward-facing truncate spines; fruit globose; seed subglobose, covered in polygonal rounded depressions, deeply scalloped on one side: *L. opacum* (page 35)

Leaflet margin unarmed; fruit ovoid; seed ovoid, flattened, with a linear depression on one side: *L. laeve* (page 38)

Cirrus armed with recurved spines, acanthophylls much reduced or absent; leaflets lanceolate: *L. korupensis* (page 42)

Canes robust, stems with sheaths >3.5 cm in diameter, with >12 leaflets on each side of the rachis; common in forest gaps and open areas:

Petiole on mature stems <20 cm long; leaflets linear-lanceolate:

Sheaths lightly to moderately armed; petiole, rachis and cirrus conspicuously light green to yellow, ocrea short (usually 12–20 cm) truncate, rounded; leaflets mid- to dark green, especially on upper surface, held ± horizontal to rachis; rachis bracts with conspicuous yellow band at base: *L. acutiflorum* (page 44)

Sheaths moderately to profusely armed; petiole, rachis and cirrus mid to dark green; ocrea usually long (20–30 cm), gradually tapering; leaflets glaucous blue-green, conspicuously pendulous on rachis; rachis bracts dry throughout: *L. robustum* (page 46)

Petiole on mature stems >20 cm long; leaflets sigmoid, elongate, held horizontally to rachis: *L. secundiflorum* (page 48)

Laccosperma opacum

(G. Mann & H. Wendl.) Drude

Botanische Zeitung 35: 635 (1877)

(Latin) "darkened" or "dull"; refers to dark green leaflets

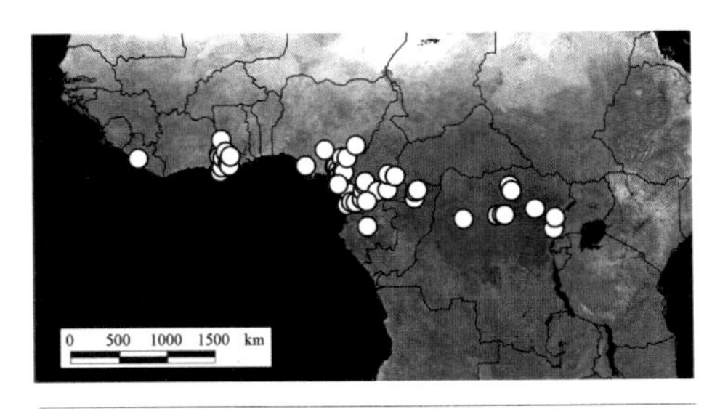

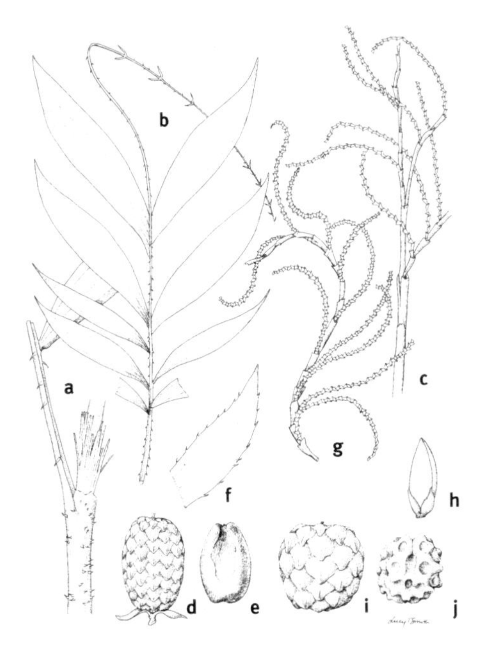

(**a.**–**e.** refer to *L. laeve*); **f.** Leaflet section; **g.** Inflorescence; **h.** Flower in bud; **i.** Fruit; **j.** Seed.

Description

An understorey **hapaxanthic** rattan palm with conspicuous globose fruits and warty seeds. **Clustering** slender rattan. **Stems** branching, 10–15 m long, up to 15 mm in diameter. **Sheath** 20 mm in diameter sparsely to moderately armed with black-tipped, upward pointing spines; **ocrea** 12–30 cm long, dry, papery and tattering, tapering at the apex, dark shiny brown within. **Leaves** petiolate, up to 50–70 cm long; **leaflets** dark green, sigmoid, composed of 2–4 folds, armed along the margin with stout spines; **cirrus** up to 50 cm; **acanthophylls** up to 2.5 cm long. **Inflorescences** produced simultaneously in the distal 30–50 cm portion of stem, rachis branches up to 50 cm in length, rachillae 10 cm long, irregularly wavy, rachis bracts striate. **Fruit** rounded, covered with large scales; **seed** 6–8 mm in diameter covered in rounded depressions, deeply scalloped on one side; **sarcotesta** white, 1 mm thick.

Uses

This species is rarely used for basketry or weaving although the stems are cut and the potable sap drunk, and the palm heart may be roasted and eaten.

Conservation status

Not threatened.

Habitat Tolerant of deep shade, *L. opacum* is commonly found in the understorey of high forest. Prefers well-drained soil and is the only species of rattan found on basalt and other volcanic soil types.

Distribution Widely distributed from Sierra Leone to Cameroon, Gabon and Congo and eastwards across the Congo Basin.

Habit in cultivation

Leaf sheath and ocrea

Leaflets (note spines on margin)

Aggregated inflorescence

Habit in forest

Aggregated infructescence

Fruit

Vernacular names

Ghana: *eholobaka* (Nzema); *sayai* (Akan-Asanti); *edem* (Kwawu): **Nigeria:** *abu* (Edo); ekwe o*ya* = cane for tie-tie (Igbo). **Cameroon:** *liko ko'ko* = "close to cane" (Mokpwe); *ge- nomé-echié* = "slave to cane rope" (Denya). **Equatorial Guinea:** *npue-nkan* (Fang). **Gabon:** *ibulu* (Myene); *di-bulu* (Sira); *di-bulu* (Lumbu); *abulo* (Kele); *éboa* (Tsogo) *ulóngó-mwa-iki* (Benga). **Congo:** *kimbana ki mukaana* (Téké).

Laccosperma laeve

(G. Mann & H. Wendl.) H. Wendl.

Les Palmiers 249 (1878)

(Latin) "smooth"; refers to the seed coat

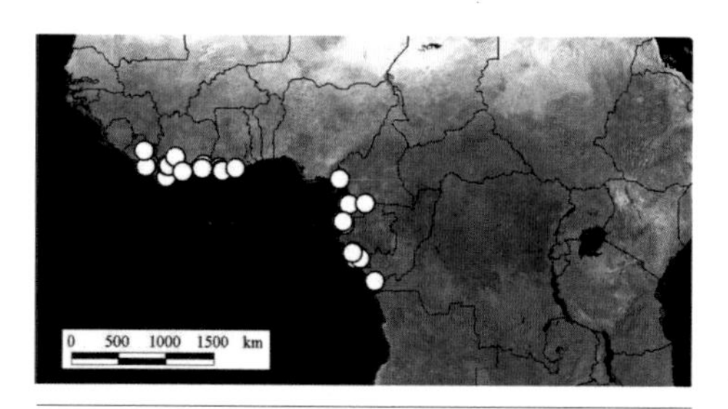

Description

A forest understorey rattan palm with spineless leaflet margins and coffee-bean like seeds. **Clustering** slender rattan. **Stems** often branching, 10–13 m long, up to 16 mm in diameter. **Sheath** 20 mm in diameter sparsely to moderately armed with brown, black-tipped spines; **ocrea** 8–20 cm long, dry, tapering at the apex, glossy mid-brown within. **Leaves** 60–90 cm long; **leaflets** dark green, narrowly sigmoid, composed of 2–4 folds, margins unarmed; **cirrus** up to 70 cm; **acanthophylls** 2.5–2.8 cm long. **Inflorescences** produced simultaneously in the distal 30–50 cm portion of stem, rachis branches erect, up to 30 cm in length, rachillae 8–12 cm long. **Fruit** ovoid, covered with fine scales; **seed** shaped like a coffee bean, flattened on one side, with a light linear depression running from base to apex; **sarcotesta** absent.

Uses

Rarely used for basketry or weaving although in Gabon the stems are pounded and woven into a rope. In the Congo the roots are roasted and eaten to improve virility.

Conservation status

Not threatened.

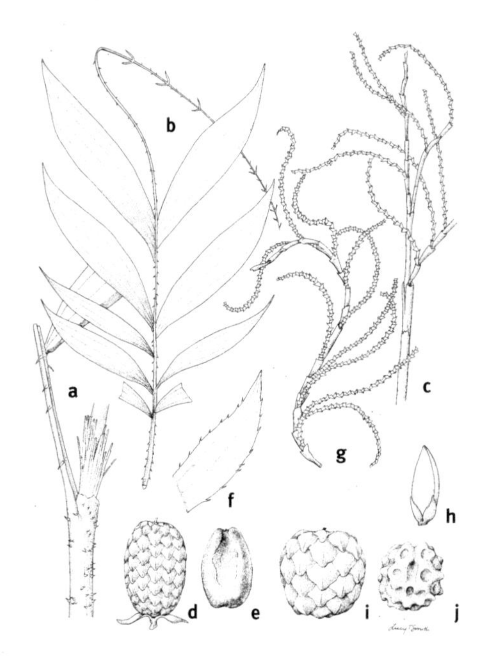

a. Sheath; **b.** Leaf; **c.** Inflorescence; **d.** Fruit; **e.** Seed; (**f.–j.** refer to *L. opacum*).

Habitat Commonly found in the understorey of high forest, particularly on seasonally-inundated soils.

Distribution In coastal forests from Liberia and the forests of Upper Guinea to Cameroon and southwards to Cabinda.

Laccosperma laeve

Habit (juvenile with spear leaf)

Habit (adult)

Leaflet detail (note lack of spines on margin)

Off-shoot causing branching of stem

Infructescence

Fruit and seed (note smooth seed coat)

Leaf

Vernacular names

Côte d'Ivoire: *ailé-mla* (Anyin). **Ghana:** *nguni* (Wasa). *tenan muhunu* = "it lives in the world for nothing" (Twi). **Nigeria:** *itunibia* (Ekit). **Cameroon:** *ge- nomé-echié* = "slave to cane rope" (Denya). **Central African Republic:** *gao* (Banda-Yangeri). **Equatorial Guinea:** *ndele* (Fang). **Gabon:** *munyengi* (Sira); *tèkè* (Tsogo).

Laccosperma korupensis

Sunderl.

Kew Bulletin 58: 987–990 (2003)

Named after Korup National Park, Cameroon, where this species is particularly abundant

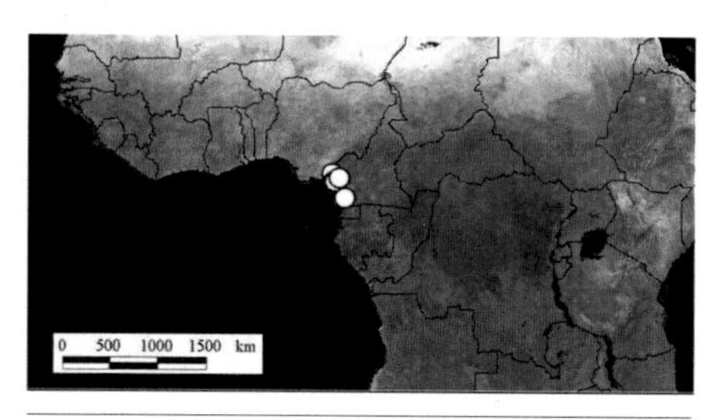

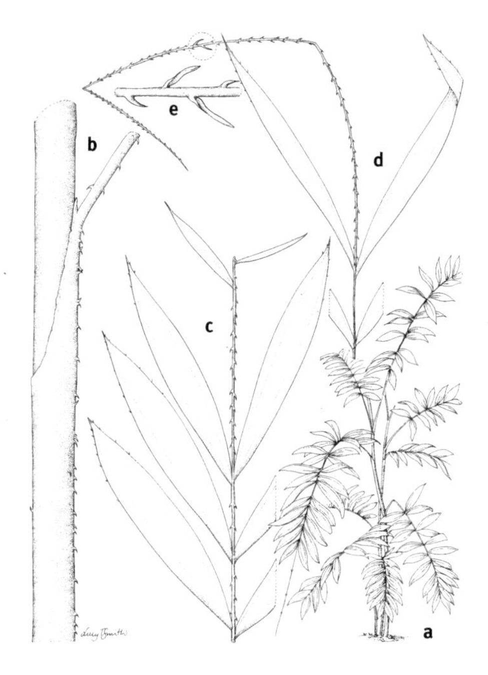

a. Habit; **b.** Stem; **c.** Leaflets; **d.** Leaflet detail; **e.** Acanthophylls.

Description
Similar in appearance to *Laccosperma laeve* and *L. opacum* but characterised by the possession of a leaf sheath that is oval in cross section and a cirrus curiously devoid of prominent acanthophylls. **Clustering** slender rattan. **Stems** up to 10 m long, up to 12 mm in diameter. **Sheath** 15 mm in diameter, oval in cross section, moderately to very sparsely armed with fine, rather brittle, spines; sheath is often unarmed near leaf junction; **ocrea** 7–10 cm long, dry, gradually tapering at the apex, often tattering and disintegrating, shiny dark brown within. **Leaves** petiolate, up to 1 m long; **leaflets** linear-lanceolate, often finely acuminate at apex, conspicuously pendulous, armed along margins with small reflexed spines; **cirrus** up to 70 cm long; **acanthophylls** much reduced or completely absent. **Flowers** unknown. **Fruit** unknown.

Uses
None known.

Conservation status
Vulnerable due to its localised and restricted range.

Habitat As with the other small-diameter canes within the genus, this is a species of the forest understorey.

Distribution From SW Cameroon, with an outlier in southern Cameroon.

Habit

Mature sheath

Habit

Leaflets

Leaf and cirrus (note reduced acanthophylls)

Juvenile sheath

Vernacular names

None recorded.

Laccosperma acutiflorum

(Becc.) J. Dransf.

Kew Bulletin 37(3): 456 (1982)
(Latin) refers to acuminate calyx lobes

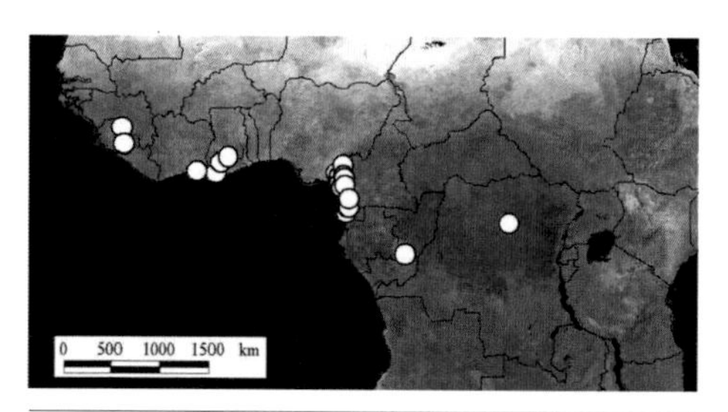

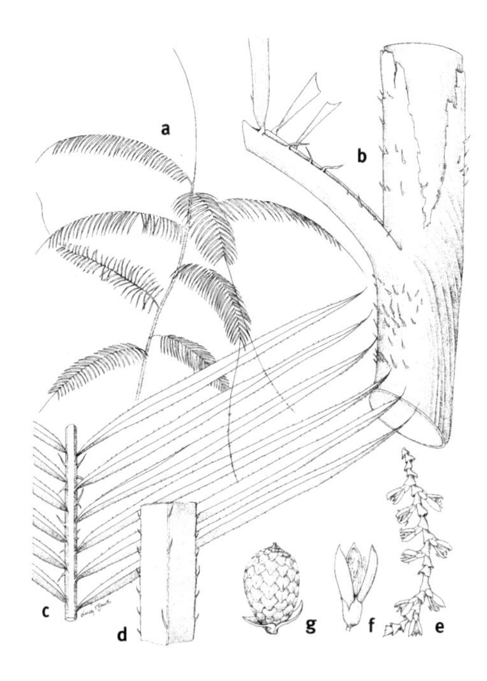

a. Habit; **b.** Stem; **c.** Leaflets; **d.** Leaflet section (underside); **e.** Flowers on rachilla; **f.** Flower; **g.** Fruit.

Description

Clustering robust to massive **hapaxanthic** rattan. **Stems** up to 70 m long, more commonly 30–50 m, 35–60 mm in diameter. **Sheath** often bright yellowish-green, 45–80 mm in diameter, sparsely to moderately armed with short, angular, black-tipped spines; **ocrea** short, 12–20 cm long, dry, tapering at the apex to form a rounded lobe, sometimes splitting longitudinally, crimson-brown within. **Leaves** shortly petiolate, up to 1.5–2 m long; **leaflets** mid-to dark green, linear-lanceolate, bluntly acuminate to apiculate at apex, held horizontally, armed along the margin with short, robust spines; **cirrus** 1.8–2.5 m long; **acanthophylls** up to 5 cm long. **Inflorescences** produced simultaneously in the distal 1.5–2.5 m portion of stem, rachis branches up to 50 cm in length, perpendicular to main axis; rachillae 20–30 cm long, pendulous, rachis bracts bright yellow-green. Hermaphrodite **flowers** in pairs, corolla lobes characterised by triangular to acuminate apex.

Note: previously confused with *L. secundiflorum,* this species is easily identified in the field by its yellow-green leaf sheaths, its robust size and horizontally-held leaflets.

Uses

Despite the large stem size, this species is reported to possess very poor quality cane and therefore is rarely used.

Conservation status

Not threatened.

Habitat A light demanding species commonly found in gap vegetation and in open areas. Responds well to selective logging and will colonise recently disturbed soil particularly on skid trails and roadsides.

Distribution Widely distributed from Sierra Leone to Cameroon, southwards to the Congo Basin. Despite its wide distribution, this species is particularly poorly represented in herbaria.

Laccosperma acutiflorum Juvenile Habit

Infructescence Leaf sheath and ocrea

Flowers Fruit

Vernacular names

Nigeria: *ukpekpe* (Ekit). **Cameroon:** *giant cane* (Pidgin). **Equatorial Guinea:** *ekwass* (Fang).

Laccosperma robustum

(Burr.) J. Dransf.

Kew Bulletin 37(3): 457 (1982)

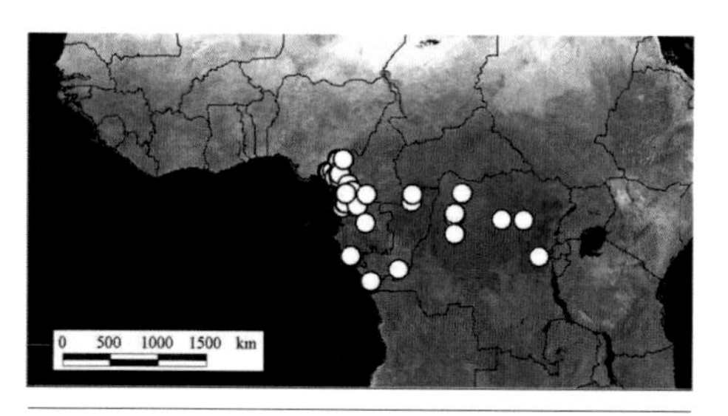

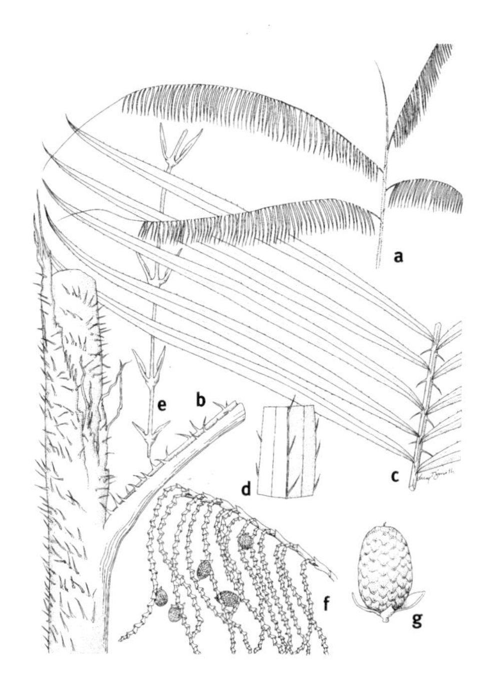

a. Habit; **b.** Stem; **c.** Leaflets; **d.** Leaflet detail; **e.** Acanthophylls; **f.** Infructescence; **g.** Fruit.

Description

A robust species of rattan cane characterised by highly pendulous leaflets. **Clustering** moderately- sized to highly robust **hapaxanthic** rattan. **Stems** 30–45 m long, 30–50 mm in diameter. **Sheath** 45–60 mm in diameter moderately to profusely armed with fine spines; **ocrea** 20–30 cm long, dry, gradually tapering at the apex, often tattering and disintegrating, crimson-brown within. **Leaves** shortly petiolate, up to 1–1.5 m long; **leaflets** glaucous blue-green, linear-lanceolate, often finely acuminate at apex, conspicuously pendulous, armed along margins with very slender hair-like spines; **cirrus** 1.5–2 m long; **acanthophylls** up to 4 cm long. **Inflorescences** produced simultaneously in the distal 1.5–2 m portion of stem, rachis branches up to 50 cm in length, perpendicular to main axis; rachillae 18–25 cm long, pendulous. **Flowers** hermaphrodite, in pairs.

Uses

Highly prized as a source of cane and widely traded throughout its range. The whole stems are used predominantly for furniture frames or are split and woven for coarse basketry.

Conservation status

Not threatened.

Habitat More commonly occurring in secondary forest, often found in forest gaps, regrowth vegetation and on roadsides. It is encountered in both terra firma and seasonally inundated forest.

Distribution From SE Nigeria and Cameroon, south to Cabinda and westwards into the Congo Basin.

Laccosperma robustum

Habit

Infructescence

Leaflets

Fruits

Vernacular names

Cameroon: *eka* (Ewondo); *nkan, aka* = cleaned cane (Bulu); *dikah* (indef.) *mekah* (def.) (Bakundu-Balue); *gekwiya* (Denya); *makak* (Trade). **Central African Republic:** *gao* (Banda-Yangere). **Equatorial Guinea:** *nkan, aka* = cleaned cane (Fang). **Gabon:** *asperge* (nom forestier). **DR Congo:** *ekpale-ekpale* (Bwa): *li-sele* (Lombo); *nkao* (Ngbaka-Ma'bo); *ikoonga* (Lombo).

Laccosperma secundiflorum

(P. Beauv.) Küntze

Revision Genera Plantarum 2: 729 (1891)

(Latin) refers to the inflorescence structure with hermaphrodite flowers in pairs

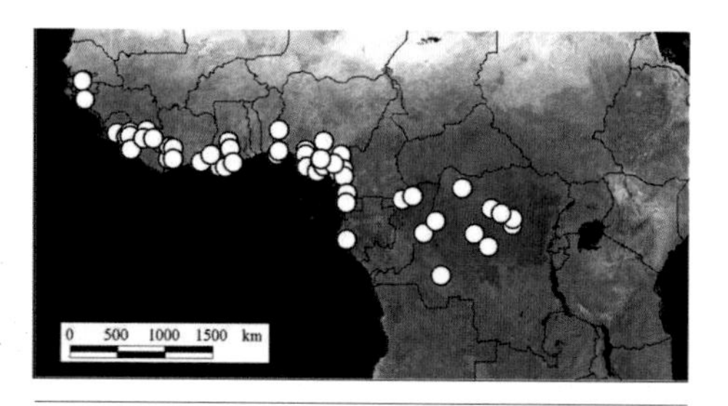

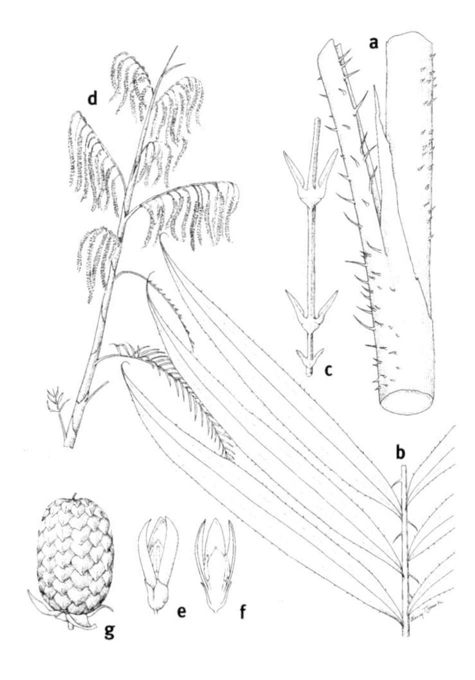

a. Sheath; **b.** Leaflets; **c.** Acanthophylls; **d.** Inflorescence; **e.** Flower; **f.** Flower section; **g.** Fruit.

Description

A robust species of rattan characterised by wide leaflets, more commonly found in high forest. **Clustering** moderately-sized to robust **hapaxanthic** rattan. **Stems** 25–50 m long, 30–35 mm in diameter. **Sheath** 20–25 mm in diameter sparsely to moderately armed with fine spines; **ocrea** 25–35 cm long, dry, often tattering, tapering at the apex, dark brown or maroon within. **Leaves** petiolate, up to 1.2–1.5 m long; **leaflets** composed of 2–4 folds, sigmoid, dark green; **cirrus** 1.5–1.8 m long; **acanthophylls** up to 4 cm long. **Inflorescences** produced simultaneously in the distal 1–1.8 m portion of stem, rachis branches 25–30 cm long, perpendicular to main axis; rachillae 15–20 cm long, pendulous. **Flowers** hermaphrodite, in pairs or less commonly in 3's.

Uses

Highly prized as a source of cane and widely traded throughout its range. Used predominantly for furniture frames and coarse basketry.

Conservation status

Not threatened.

Habitat High forest, occurring rarely in secondary forest, commonly found under a forest canopy. Particularly abundant in seasonally inundated and swampy areas.

Distribution From Sénégal in West Africa, across the Upper Guinea forests to Cameroon, south throughout the Congo Basin.

Habit

Leaflets

Sheath and petiole

Ocrea

Aggregate (hapaxanthic) inflorescence

Acanthophylls

Vernacular names

Senegal: *ka-likut* (Jola-Fogny). **Guinea-Bissau:** *tambem-hadje* (Fulfulde-Pulaar); *tambendjom* (indef.), *tambendjom-ô* (def.) (Mandinka). **Sierra Leone:** *lumboinyo-piando* (Kisi); *kangane* (Kono); *kafo* (Loko); *kavo* (def. *kavui*) (Mende); *ka-gbesu* = whole stems, *e-gbak* = leafless part of the stem (Themne). **Côte d'Ivoire:** *kumh* (Attié); *agué* (Ebrié); *djoho, djolo* (Krumen); *ahika* (Anyin); *gblé* (Godié). **Ghana:** *willow* (Trade); *ayié* (Akan-Asanti); *ayike* = large rattan (Nzema). **Benin:** *kpanon* (Defi); *kpacha* (Gun-Gbe). **Nigeria:** *ohwara* (Urhobo); *okankan* = whole cane, *ukwen* = when split (Edo); òbóng (Efik); *ukpé* = cane rope made of this species (Ijo-Izon); *iga* (Ekpeye); *añà* (Igbo); *epe-nla, ikan-ikó* = a hook (Yoruba). **Cameroon:** *ka-kawa* (Baka); *ekwos* (Balundu-Bima); *nde-gekwiya* (Denya). **Gabon:** *nkan* (Fang); *nkanda* (Kélé); *ikandji* (Kota); *okana* (Ndumu); *mokangé* (Pinji); *mokangé* (Tsogo); *mukanda* (Sira); *mukanda* (Duma); *mukanda* (Lumbu); *nkogu* (Myene); *nkanyi* (Seki). **Congo:** *mukaana a nguomo* (Téké). **DR Congo:** *ma-kauw, bo-kauw* (def.) (Lingala); *bo-nganga* (Mongo-Nkundu); *nkau* (Kongo). **Angola:** *mi-cau* (Mbundu-Luanda).

Key to the species of *Oncocalamus*

Canes moderate to robust; stems with sheaths >20 mm in diameter; ocrea truncate with ± conspicuous 1.5–2.5 cm long rounded lobe; leaves, including cirrus, >1 m long, leaflets linear-lanceolate or only very mildly sigmoid:

Stems with sheaths <30 mm in diameter, moderately armed, with no swelling beneath the leaf; leaf with 25–35 pairs of leaflets on each side of the rachis, inflorescence ± 1 m long, flower cluster with variable number of pistillate flowers (1, 3, 5, or 7) in each, seeds with rounded polygonal depressions, warty: *O. mannii* (page 51)

Stems with sheaths >30 mm in diameter, unarmed or moderately to profusely armed, with visible rounded swelling beneath leaf; leaf with >35 pairs of leaflets on each side of the rachis; inflorescence >1 m long; flower clusters always 3 pistillate flowers, seeds smooth:

Leaf sheath moderately to profusely armed, ocrea almost horizontal, truncate, with small (<1 cm) rounded lobe abaxial to the leaf; inflorescence up to 1.2 m long; peduncular and rachis bracts <8 cm long; rachillae bright yellow: *O. macrospathus* (page 54)

Leaf sheath barely armed or unarmed, ocrea with high (>1 cm) rounded lobe abaxial to the leaf; inflorescence up to 1.8 m long; peduncular and rachis bracts >8 cm long; rachillae deep crimson: *O. tuleyi* (page 57)

Canes slender; stems with sheaths <20 mm in diameter; ocrea horizontally truncate, extending to 3 cm beyond leaf; leaves, including cirrus, <1 m in length; leaflets sigmoid, composed of 2–4 folds: *O. wrightianus* (page 60)

Oncocalamus mannii

(H. Wendl.) H. Wendl.

Les Palmiers 244 (1878)
Gustav Mann (1836–1916), German botanist and horticulturist

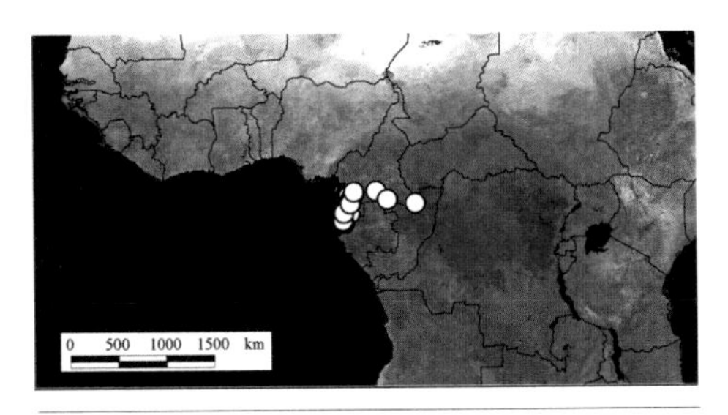

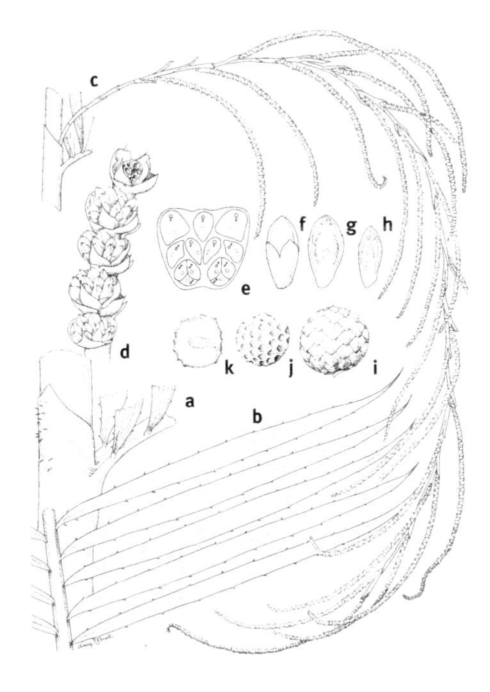

a. Stem; **b.** Leaflets; **c.** Inflorescence; **d.** Portion of rachilla; **e.** Flower cluster diagram; **f.** Pistillate flower; **g.** Pistillate flower section; **h.** Staminate flower; **i.** Fruit; **j.** Seed; **k.** Seed section.

Description

A slender to moderately-sized rattan cane characterised by the presence of a variable number of pistillate flowers in a cluster (3–7) and conspicuously warty seeds. **Clustering pleonanthic, monoecious** rattan palm. **Stems** 15–30 m long, 8–16 mm in diameter. **Sheath** 12–28 mm in diameter sparsely to moderately armed with triangular brown-black spines, particularly concentrated on apex of sheath and ocrea; **ocrea** truncate, saddle shaped, up to 2 cm long. **Juvenile leaves** sessile, strongly **bifid**. **Elaminate rachis** often present, particularly at base of stem. **Adult leaves** sessile, or shortly petiolate, 1–1.2 m long; **leaflets** linear-lanceolate to mildly sigmoid, lightly pendulous; **cirrus** 1–1.5 m long; **acanthophylls** up to 2.5 cm long, green-brown. **Inflorescences** produced in successive axils 3–5 m from apex, pendulous, up to 1.5 m long, rachillae and rachis bracts at first bright crimson becoming dry, brown. **Flower** cluster with 1–3 central pistillate flowers, 2 further cincinni of 1–2 pistillate and 2–3 staminate flowers. **Seed** covered with polygonal depressions to give warty appearance; **sarcotesta** white, <0.5 mm thick.

Uses

The cane of this species is poor in quality; it is rather inflexible and prone to breaking. However, particularly in the absence of other species, *O. mannii* can be employed for coarse basketry.

Conservation status

Not threatened.

Habitat A light demanding species common in forest gaps and other open areas.

Distribution Restricted from southern Cameroon to Gabon.

Oncocalamus mannii

Juvenile in cultivation

Habit

Inflorescence

Infructescence

Fruit

Leaf sheath

Vernacular names

Cameroon: *mfop n'lon* (Bulu). **Equatorial Guinea:** *asa-nlong* (juvenile), *ndoro* (adult) (Fang). **Congo:** *mituo* (Téké).

Oncocalamus macrospathus

Burr.

Notizblatt Botanische Garten Museum Berlin-Dahlem 15: 749 (1942)

(Latin) large bracts on inflorescence

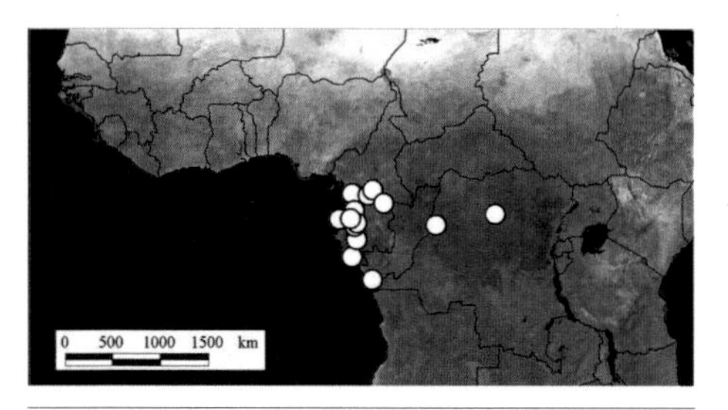

Description

A robust rattan with bright yellow rachillae on the inflorescence, 3 central pistillate flowers and a smooth seed coat. **Clustering pleonanthic, monoecious** rattan palm. **Stems** 20–35 m long, 18–30 mm in diameter. **Sheath** 28–40 mm in diameter, moderately to profusely armed with upward-pointing triangular brown spines, particularly concentrated on apex of sheath and ocrea; **ocrea** striate, truncate, sometimes with slight lobe, up to 1 cm long. **Juvenile leaves** sessile, **bifid**, soon becoming pinnate. **Elaminate rachis** often present, particularly at base of stem, up to 2 m long. **Adult leaves** sessile, or very shortly petiolate, 1.5–2 m long; **leaflets** linear-lanceolate, arching, pendulous, armed with robust spines along margin; **cirrus** 1.2–1.5 m long; **acanthophylls** up to 5 cm long, yellow-green. **Inflorescences** produced in successive axils 3 m from apex, pendulous, up to 1.7–2 m long, rachillae and rachis bracts at first bright yellow becoming dry, brown. **Flower** cluster with 1 central pistillate flower with 2 further cincinni of 1 pistillate and 2–3 staminate flowers. **Seed** coat smooth; **sarcotesta** white, <0.3 mm thick.

Uses

None recorded.

Conservation status

Not threatened.

a. Sheath; **b.** Leaflets; **c.** Inflorescence; **d.** Portion of immature rachilla; **e.** Flower cluster diagram; **f.** Pistillate flower with prophyllar bract; **g.** Pistillate flower section; **h.** Staminate flower section; **i.** Fruit on rachilla; **j.** Fruit; **k.** Seed.

Habitat Commonly found in forest margins, tree-fall gaps and other open areas and particularly common in seasonally-inundated forest and alongside water courses.

Distribution From Cameroon, south of the Sanaga River, to Cabinda (Angola) and across the Congo Basin.

Oncocalamus macrospathus

Habit

Juvenile

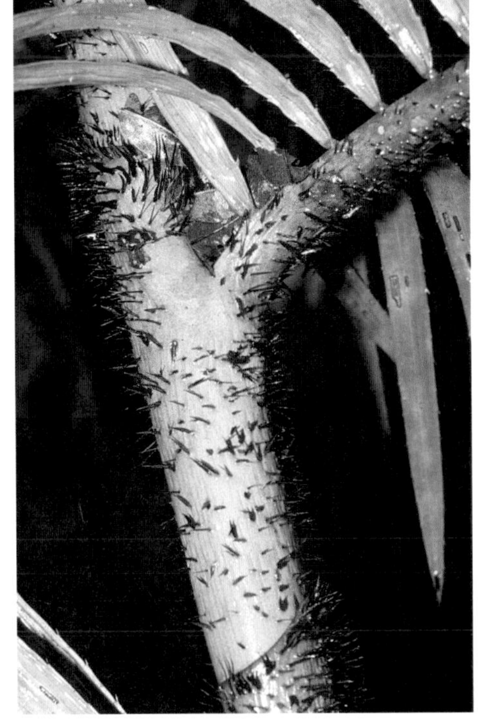

Leaf sheath

Inflorescence

Infructescence

Fruit

Vernacular names

Cameroon: *eboti* (Ewondo).

Oncocalamus tuleyi

Sunderl.

Journal of Bamboo & Rattan 1(4): 361–369 (2002)

Paul Tuley (1929–2004) agronomist and civil servant

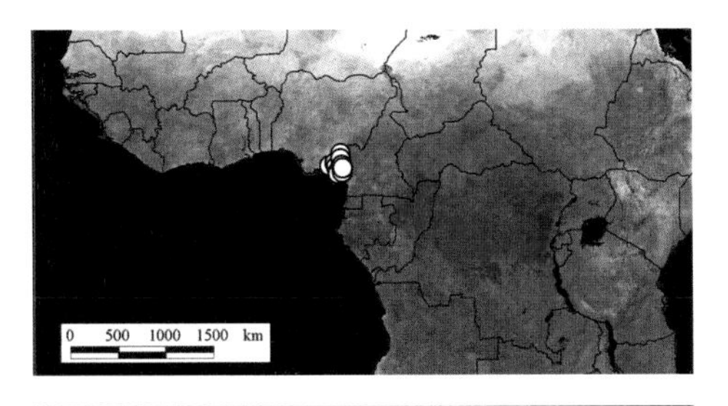

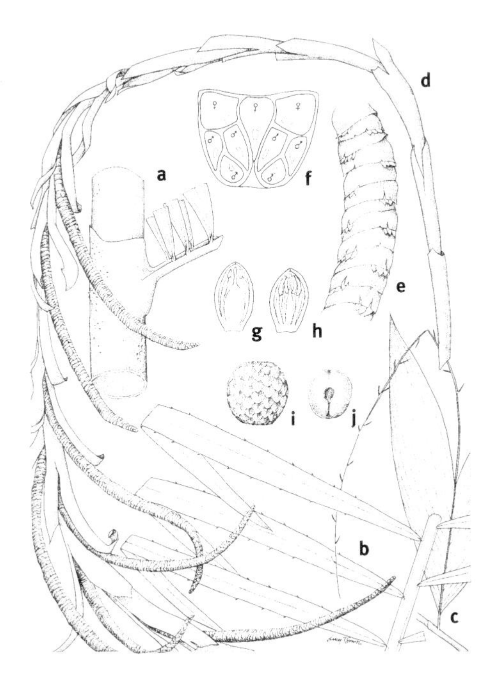

a. Sheath; **b.** Leaflets; **c.** Juvenile leaf; **d.** Inflorescence; **e.** Portion of immature rachilla; **f.** Flower cluster diagram; **g.** Staminate flower; **h.** Pistallate flower; **i.** Fruit; **j.** Seed.

Description

A robust cane with relatively bare sheaths, dull crimson rachillae, 3 central pistallte flowers and a smooth seed coat. **Clustering pleonanthic, monoecious** rattan palm. **Stems** up to 30 m long, rarely to 50 m, 13–22 mm in diameter. **Sheath** 25–45 mm in diameter, sparsely to patchily armed with dark brown to glaucous black spines, often concentrated on the ocrea; spines often sloughing off to leave raised blister-like scars; **ocrea** saddle-shaped, extending to 2.5 cm. **Juvenile leaves** shortly petiolate, **bifid**, soon becoming pinnate. **Elaminate rachis** often present, particularly at base of stem, up to 2 m long. **Adult leaves** sessile, or very shortly petiolate, 1.2–2 m long; **leaflets** linear-lanceolate, arching, pendulous, **cirrus** 0.8–1.5 m long; **acanthophylls** up to 4 cm long, crimson-brown. **Inflorescences** produced in successive axils 3 m from apex, pendulous, up to 1.8 m long, rachillae and rachis bracts flattened, at first dull crimson, becoming dry, brown. **Flower** cluster with 1 central pistillate flower with 2 further cincinni of 1 pistillate and 3–4 staminate flowers. **Seed** coat smooth, with linear cleft or rounded depression below; **sarcotesta** white, <0.3 mm thick.

Uses

Not used commercially due to the poor quality of the cane; it is rather weak, inflexible and prone to breakage on bending. However, the base of the leaf sheath is often used by the indigenous communities of SW Cameroon as a chew-stick. In Nigeria, the stem epidermis is often used for tying yams to their climbing poles ("yam-barns") as it rots relatively quickly and does not constrict the developing stems.

Conservation status

Vulnerable.

Oncocalamus tuleyi

Habit

Elaminate rachis

Leaf sheath

Juvenile in forest

Inflorescence showing large bracts

Juvenile bifid leaf

Habitat Forest edge, adjacent to open areas, and in gap regrowth vegetation in forest. An early coloniser of disturbed land and as such a characteristic feature of roadside vegetation in logged forest.

Distribution Restricted to the coastal forests of SE Nigeria to SW Cameroon, north of the Sanaga River.

Vernacular names

Nigeria: *iboh* (Ekit). **Cameroon:** *madame* (Trade/Pidgin); *mo'ap* (Balundu-Bima); *edju* (Bakundu-Balue); *moa-echié* (Denya).

Oncocalamus wrightianus

Hutch.

Flora of West Tropical Africa 2: 391 (1936)
Charles Henry Wright (1864–1941), British botanist

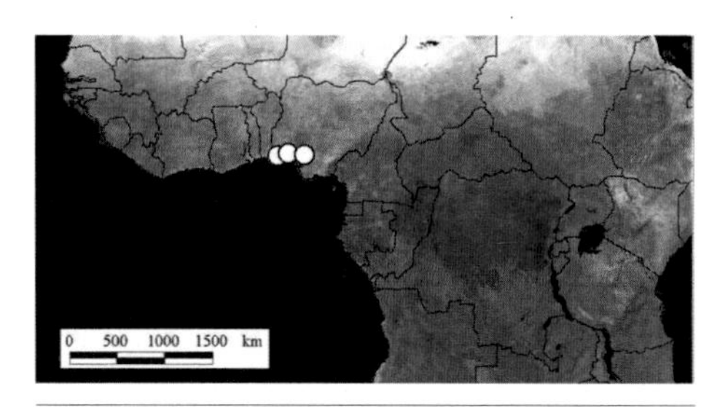

Description

A fine, slender rattan with somewhat sigmoid leaflets. **Clustering** slender rattan palm. **Stems** up to 10 m long, 6–10 mm in diameter. **Sheath** 8–155 mm in diameter, sparsely to moderately armed with caducous black spines, particularly concentrated on the ocrea; spines often sloughing off to leave raised, triangular, blister-like scars; **ocrea** horizontally truncate, extending to 3 cm. **Leaves** sessile, or very shortly petiolate, 60 cm long; **leaflets** composed of 2–4 folds, broadly lanceolate or ovate when juvenile, sigmoid at maturity; **cirrus** 35–45 cm long; **acanthophylls** up to 1–1.8 cm long. **Flowers** unknown. **Fruit** unknown.

Uses

In Nigeria the split stem is used to make tying materials of different sorts: coarse cordage, fine twine, and string or thread. The leaf sheath base is also used by the Igbo as a chew stick.

Conservation status

Endangered.

Sheath and leaf section

Habit

Habitat Recorded from swamp forest and the savannah forest margin.

Distribution Known only from southern Benin and Nigeria, extending to the Niger Delta.

Vernacular names

Benin: *hofle* (Defi); *gbe-dekun* (Gun-Gbe).
Nigeria: *akwal´* (Igbo); *pankéré* (Yoruba).

Cane bridge, Nyang, Cameroon

Making farm baskets (kenjas), Mone, Cameroon

Cassave sifter, Takamanda, Cameroon

Farm baskets, Banga-Bakundu, Cameroon

Fabricated chairs, Bata, Equatorial Guinea

Calamus deërratus

G. Mann & H. Wendl.

Philosophical Transactions of the Linnean Society 24: 429 (1864)

(Latin) "to go astray", which refers to the habit of the species to form expansive clumps

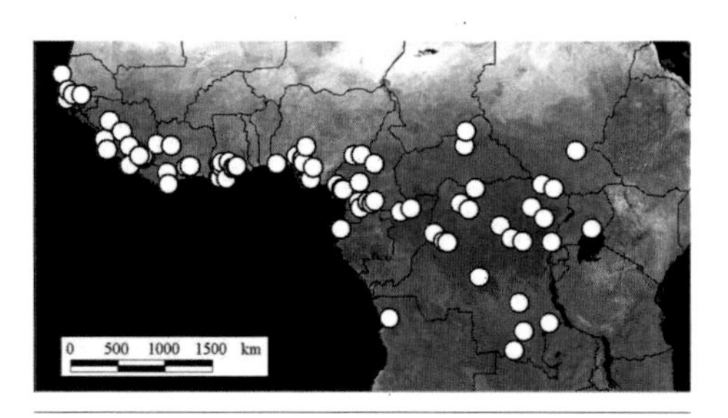

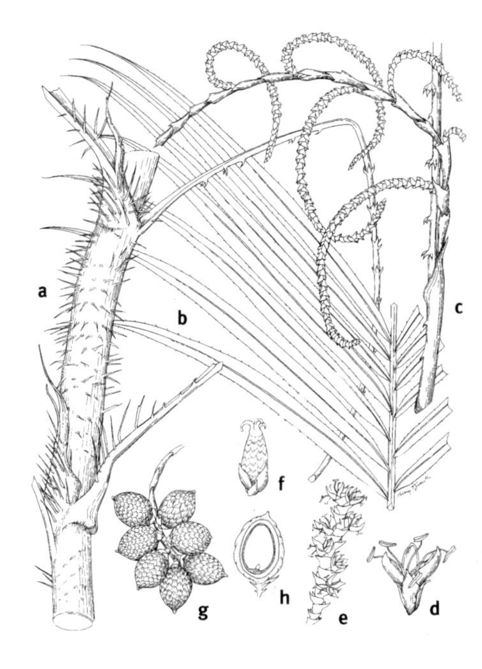

a. Stem; **b.** Leaflet section; **c.** Inflorescence; **d.** Male flower; **e.** Male rachilla; **f.** Female flower; **g.** Fruit; **h.** Fruit section.

Description

A rattan of swampy forests often forming dense monospecific stands. **Clustering** rattan forming dense clumps. Climbs by means of **flagellum,** arising directly from the sheath. **Stems** erect, prostrate or scandent, up to 20 m in length, 1–2.8 cm in diameter. **Sheath** 1.2–3.5 cm in diameter, armed with broadly triangular brown-black spines up to 3 cm in length; **ocrea** 8–10 cm long, papery, often drying with spines concentrated on the margins to form a distinct cleft; **knee** present below leaf, horizontal, somewhat folded. **Leaves** 1.2–1.5 m long; **leaflets** often grouped in 3's to 6's, particularly on juvenile leaves. **Flowers** present on flagellum, with both male and female flowers occurring on different individuals (**dioecious**). **Fruit** globose, somewhat beaked at apex.

Uses

Widely used as a source of cane, particularly in the absence of more desirable species. Commonly used for weaving, although can be used for furniture frames if 2–3 stems are joined together. Traded in West Africa (Gambia and Senegal to Ghana), southern Africa (Zambia, Zimbabwe) and East Africa (Uganda, Kenya and Tanzania). Not traded in Central Africa as the cane is considered inferior to other species of higher quality.

Conservation status

Not threatened.

Habitat Commonly found in swamp forest and riverine areas as well as gallery forest in savannah woodland. Can occur up to altitudes of 1500 m particularly in East Africa, although more commonly occurs at <500 m.

Distribution Across the humid forest zone of Africa from the Gambia, southwards to northern Angola and Zambia and eastwards to southern Sudan, Uganda and Tanzania.

Ocrea

Knee

Habit

Male inflorescences on flagellum

Male inflorescence on sheath

Vernacular names

Senegal: *ki tid* (Balanta); *kintem* (Bainouk); *mantampa da sera* (Crioulo, Upper Guinea); *bu kètao bu ketav, fu fiaf, ka kèt, ka tay, ke hiya, kékiya* (Jola-Fogny); *tambem* (Fula-Pulaar); *tambi* (Tukulor); *tambo* (Mandinka); *tãbi* (Malinke); *e kapat* (Mandyak); *ratlan* (Wolof). **Gambia:** *tambo* (Mandinka). **Guinea-Bissau:** *quitite* (Balanta); *batanou* (Biafada); *mantampa de sera* (Crioulo, Upper Guinea); *tambem* (Fulfulde-Pulaar); *tambo* (Mandinka); *ecapate* (Mandyak); *quito* (Papel). **Guinea:** *tambo* (Mandinka); *tâbi* (Malinke). **Sierra Leone:** *lumboinyo-lando* (Kisi); *kanga-mese* (Kono); *tambe* (Loko); *tambi* (Maninka); *tamba* (def. *tembuî*) (Mende); *tambi* (Susu); *ra-gbet* (Themne); *tambu-na* (Yalunka). **Liberia:** *kpa kala* (Mano). **Côte d'Ivoire:** *ailé-mlé* (Anyin); *gapapa* (Godié). **Ghana:** *demmeré* (Twi, also trade name); *néné,* (Akan); *ayeka* (Anufo); *ayeka* (Sehwi); *keteku* (Éwé); *ayeké* (Nzema). **Benin:** *akete* (Defi); *dekun wéwé* (Gun-Gbe). **Nigeria:** *erogbo, erugbo* (Edo); *ekwe-oji, iye* (Igbo); *apié* (the plant itself, or the cane-rope made from it) (Ijo-Izon), *bwálãm* (a cane) (Pero); *erogbo, erugbo* (Yoruba). **Cameroon:** *nding* (Bulu). **Equatorial Guinea:** *nzing* (Fang). **Central African Republic:** *bioh* (Banda-Yangere). **DR Congo:** *kpude* (Zande); *ma-ndakele* (Ngbaka-Ma'bo); *ikonga* (Lombo); *babio* (Mongo-Nkundu); *lekwe* (BaMbuti). **Uganda:** *bi-lekwe* (Amba).

CURRENTLY RECOGNISED NAMES AND SYNONYMS

CALAMUS

Calamus deërratus G. Mann & H. Wendl.
syn. *Calamus akimensis* Becc.
Calamus barteri Drude
Calamus falabensis Becc.
Calamus heudelotii Becc.
Calamus laurentii De Wild.
Calamus lepreurii Becc.
Calamus schweinfurthii Becc.

EREMOSPATHA

Eremospatha barendii Sunderland

Eremospatha cabrae (De Wild. & Th. Dur.) De Wild.
syn. *Calamus cabrae* De Wild. & Th. Dur.
Eremospatha rhomboidea Burr.
Eremospatha suborbicularis Burr.

Eremospatha cuspidata (G. Mann & H. Wendl.) H. Wendl.
syn. *Calamus (Eremospatha) cuspidatus* G. Mann & H. Wendl.

Eremospatha dransfieldii Sunderl.

Eremospatha haullevilleana De Wild.

Eremospatha hookeri (G. Mann & H. Wendl.) H. Wendl.
syn. *Calamus (Eremospatha) hookeri* G. Mann & H. Wendl.

Eremospatha laurentii De Wild.

Eremospatha macrocarpa (G. Mann & H. Wendl.) H. Wendl.
syn. *Calamus (Eremospatha) macrocarpus* G. Mann & H. Wendl.
Eremospatha sapini De Wild.

Eremospatha quinquecostulata Becc.

Eremospatha tessmanniana Becc.

Eremospatha wendlandiana Dammer ex Becc.
syn. *Eremospatha korthalsiaefolia* Becc.

LACCOSPERMA

Laccosperma acutiflorum (Becc.) J. Dransf.
syn. *Ancistrophyllum acutiflorum* Becc.

Laccosperma korupensis Sunderl.

***Laccosperma laeve* (G. Mann & H. Wendl.) H. Wendl.**
syn. *Ancistrophyllum laeve* (G. Mann & H. Wendl.) Drude
Calamus (subgen. *Laccosperma*) *laevis* G. Mann & H. Wendl.

Laccosperma opacum (G. Mann & H. Wendl.) Drude
syn. *Ancistrophyllum opacum* (G. Mann & H. Wendl.) Drude
Calamus (subgen. *Laccosperma*) *opacus* G. Mann & H. Wendl.

Laccosperma robustum (Burr.) J. Dransf.
syn. *Ancistrophyllum robustum* Burr.

Laccosperma secundiflorum (P. Beauv.) Küntze
syn. *Ancistrophyllum secundiflorum* (P. Beauv.) H. Wendl.
Calamus (subgen. *Ancistrophyllum*) *secundiflorus* G. Mann & H. Wendl.
Calamus secundiflorus P. Beauv.
***Laccosperma laurentii* (De Wild.) J. Dransf.**
Ancistrophyllum laurentii De Wild.
Ancistrophyllum majus Burr.

ONCOCALAMUS

Oncocalamus macrospathus Burr.

Oncocalamus mannii (H. Wendl.) H. Wendl.
syn. *Calamus* (*Oncocalamus*) *mannii* H. Wendl.
Oncocalamus acanthocnemis Drude
Oncocalamus phaeobalanus Burr.
Calamus niger Braun & Schum.

Oncocalamus tuleyi Sunderl.

Oncocalamus wrightianus Hutch.

FURTHER READING

Dransfield, J. 1982. Nomenclatural notes on *Laccosperma* and *Ancistrophyllum* (*Palmae: Lepidocaroideae*). *Kew Bulletin* 37(3): 455–457

Dransfield, J. 1988. The palms of Africa and their relationships. *Monographs in Systematic Botany* 25: 95–103

Dransfield, J., F.O. Tesoro & N. Manokaran (eds) 2002. *Rattan: current research issues and prospects for conservation and sustainabale development*. Food and Agriculture Organisation: Non-Wood Forest Products no. 14. Available at: www.fao.org\publishing

FAO. *Rattan glossary and compendium glossary with emphasis on Africa*. 2003. Non-Wood Forest Products no. 16. Available at: www.fao.org\publishing

Letouzey, R. 1978. Notes phytogéographiques sur les palmiers du Cameroun. *Adansonia*, sér. 2. 18(3): 293–325

Mann, G., & H.A. Wendland 1864. On the palms of western tropical Africa. *Philosophical Transactions of the Linnean Society* 24: 421–439

Moore, H.E. 1971. Wednesdays in Africa. *Principes* 15: 111–119

Morakinyo, A.B. 1995b. Profiles and pan-African Distributions of the rattan species (Calamoideae) recorded in Nigeria. *Principes* 39(4): 197–209

Profizi, J.P. 1986. Notes on West African rattans. *RIC Bulletin* 5(1): 1–3

Pyneart, L. 1911. Les palmiers utilés. *Bulletin Agricole du Congo Belge* 2: 535–552

Sunderland, T.C.H. 2002. Hapaxanthy and pleonanthy in African rattans. *Journal of Bamboo and Rattan* 1(2): 28–42

Sunderland, T.C.H. 2003a. Two new species of rattan (Palmae: Calamoideae) from Africa. *Journal of Bamboo and Rattan* 1(4): 361–369

Sunderland, T.C.H. 2003b. Two new species of rattan (Palmae: Calamoideae) from the forests of West and Central Africa. *Kew Bulletin* 58: 987–990

Sunderland, T.C.H. & J.P. Profizi (eds.) *New research on African rattans*. International Network for Bamboo and Rattan (INBAR) Proceedings no. 9. Beijing, China. Available at: www.inbar.int/publication/main.asp

Tomlinson, P.B. 1962b. Palms of Africa. *Principes*. 6: 96–103

Tuley, P. 1965. The inflorescence of Nigerian Lepidocaryoid palms. *Principes* 9: 93–98

Tuley, P. 1995. *The Palms of Africa*. Trendrine Press. UK

Unasylva (Vol. 52) 205. 2001 (special edition on rattan). Available at: www.fao.org/publishing